T0238277

Lecture Notes in Artificial Intelligence   1458

Subseries of Lecture Notes in Computer Science
Edited by J. G. Carbonell and J. Siekmann

Lecture Notes in Computer Science

Edited by G. Goos, J. Hartmanis and J. van Leeuwen

Springer
Berlin
Heidelberg
New York
Barcelona
Budapest
Hong Kong
London
Milan
Paris
Singapore
Tokyo

Vibhu O. Mittal   Holly A. Yanco
John Aronis   Richard Simpson (Eds.)

Assistive Technology and Artificial Intelligence

Applications in Robotics, User Interfaces
and Natural Language Processing

Springer

Volume Editors

Vibhu O. Mittal
Just Research and
Carnegie Mellon University
4616 Henry Street, Pittsburgh, PA 15213, USA
E-mail: mittal@justresearch.com

Holly A. Yanco
MIT Artificial Intelligence Laboratory
545 Technology Square, Cambridge, MA 02139, USA
E-mail: holly@ai.mit.edu

John Aronis
University of Pittsburgh, Department of Computer Science
Pittsburgh, PA 15260, USA
E-mail: aronis@cs.pitt.edu

Richard C. Simpson
TRACLabs, Metrica, Inc.
1012 Hercules, Houston, TX 77058, USA
E-mail: rsimpson@traclabs.com

Cataloging-in-Publication Data applied for

Die Deutsche Bibliothek - CIP-Einheitsaufnahme

Assistive technology and artificial intelligence : applications in
robotics, user interfaces and natural language processing / Vibhu O.
Mittal ... (ed.). - Berlin ; Heidelberg ; New York ; Barcelona ;
Budapest ; Hong Kong ; London ; Milan ; Paris ; Singapore ; Tokyo :
Springer, 1998
   (Lecture notes in computer science ; Vol. 1458 : Lecture notes in
   artificial intelligence)
   ISBN 3-540-64790-2

CR Subject Classification (1991): I.2, H.5.2

ISBN 3-540-64790-2 Springer-Verlag Berlin Heidelberg New York

This work is subject to copyright. All rights are reserved, whether the whole or part of the material is
concerned, specifically the rights of translation, reprinting, re-use of illustrations, recitation, broadcasting,
reproduction on microfilms or in any other way, and storage in data banks. Duplication of this publication
or parts thereof is permitted only under the provisions of the German Copyright Law of September 9, 1965,
in its current version, and permission for use must always be obtained from Springer-Verlag. Violations are
liable for prosecution under the German Copyright Law.

© Springer-Verlag Berlin Heidelberg 1998
Printed in Germany

Typesetting: Camera ready by author
SPIN 10638261   06/3142 – 5 4 3 2 1 0   Printed on acid-free paper

Foreword

In the last decade, applications of Artificial Intelligence (AI) have become common and widespread. Reports in the series of conferences on the Innovative Applications of Artificial Intelligence, for example, document successful introduction of intelligent software for a rich variety of tasks within business, finance, science, medicine, engineering, manufacturing, education, the military, law, and the arts. Applications of AI have been demonstrated to save time and money, to increase throughput, to reduce errors, and to reach better decisions. However, relatively few applications address the important problems of enhancing the quality of life for people with disabilities.

The challenges associated with designing and constructing general-purpose assistive devices are great. But the current capabilities of many AI systems closely match some of the specialized needs of disabled people. For instance, rudimentary speech recognition for a limited vocabulary may be too limiting for general use in commercial applications, but it can be immensely useful to someone with severe physical limitations. Similarly with speech generation, text understanding, language generation, limited-task manipulators, vision systems, and so on. That is, even limited solutions to carefully circumscribed problems can make a difference.

Fortunately, there is a growing interest in applying the scientific knowledge and engineering experience developed by AI researchers to the domain of assistive technology and in investigating new methods and techniques that are required within the assistive technology domain. Some areas of current work include robotic wheelchairs, and the automation of the process of converting textbooks and other written materials into recordings for the blind. It also includes new user interfaces for computers to accommodate people with different kinds and varying degrees of motor, hearing, or visual disabilities. The same kinds of AI methods and principles that achieve flexibility in current applications should be applicable to tailoring devices for specialized needs.

The domain of developing better assistive devices is particularly interesting because the interaction between the person and the system allows researchers to overcome some of the common stumbling blocks for AI applications. It seems clear, for instance, that the users of these devices will be actively engaged in trying to make them work. They may also be inclined to accept some of the limits of new devices if the quality of help provided within those limits is significant. Many assistive applications need only solve a portion of the problem that would need to be solved for a fully intelligent assistant.

While the addition of a person into the cognitive loop allows researchers in this area to avoid some of the usual difficulties, it adds a new dimension that must be considered: the user interface. Researchers in this domain must consider the specialized needs of people with disabilities, often including interviews in the research process. Assistive applications with ineffective user interfaces will be useless. Research in this area needs to build on past research on user interface technology, as well as AI, to come up with new solutions that can be tailored to the needs of specific individuals and adapt to changing needs.

Another important feature of this problem area is the cost-benefit equation. As in many medical problems, the cost of errors may be very high. Systems must be reliable when people use them in situations that are potentially dangerous, such as crossing the street. However, the benefits to those with severe disabilities are potentially as great as in any other application we can consider.

A substantial amount of AI research is clearly relevant to the applications considered here, for example, vision, locomotion, and manipulation systems for robots, planning systems, speech and text understanding and generation, decision making under uncertainty, troubleshooting systems, cognitive modeling, and intelligent interfaces. And more. But it needs careful engineering to be embedded in robust, portable, and cost-effective devices that are designed to be partners with people.

We are at a point in time when AI technology is advanced enough to make a significant difference in the lives of disabled people. The papers collected in this volume address many significant tasks and mention many more. They make it clear that researchers must understand what is needed by the disabled persons who will use the devices, as well as what is possible with the current AI technology. They also demonstrate that the benefits are not easily achieved. When the AI community is looking for meaningful challenges, these papers show us where to look.

Bruce G. Buchanan
University of Pittsburgh

Preface

This volume arose out of a long standing interest that all of the editors have had in the use and advances of technology that can help people lead a better life. The focus here is on helping users extend their current range of cognitive, sensory or motor abilities. While some readers may associate the mention of a "wheelchair" with handicapped people, it should be emphasized that this category includes *all* of us. At some point or other in our lives, we can almost certainly benefit from technology that can help us hear, speak, understand or move about more easily. As we grow older, our physical faculties may diminish and we may *need* some of these assistive devices; but even if we are in perfect health, who amongst us would not like to have a car that could navigate around obstacles by itself, or have a real-time translator translate one language to another for us in a foreign country?

To that end, workshops and symposia were organized by one or more of the editors in Montreal in August 1995 (during the International Joint Conference on Artificial Intelligence), at the Massachussetts Institute of Technology in Boston (as part of the 1996 AAAI Fall Symposium Series) and in Madison, Wisconsin (during the 1998 National Conference on Artificial Intelligence). These workshops were informal affairs and led to extensive discussions, giving rise to several fruitful collaborations and fast friends.

One of the main points underscored by the workshops was that there was a significant amount of related work being conducted in many diverse fields such as robotics, vision, planning and natural language processing. However, there was little, if any, concerted effort to bring these researchers together. The workshops mentioned previously represented sporadic efforts to meet similar researchers and inform others about this work. It was felt that a collection of papers representing some of the work in related areas might help other researchers as well as prospective graduate students looking for dissertation topics. There is a unique aspect to working in this sub-area: a human is always part of the processing loop. This introduces certain constraints which can significantly change the nature of the problem being investigated. The collection of papers in this volume illustrates some of these issues well. We hope that these papers will spark users' interest in these and related issues and think how their own research might be used for addressing similar problems. We would like to thank all of our contributors and hope you find this compendium useful in your own research.

May 1998

Vibhu Mittal
Holly Yanco
John Aronis
Rich Simpson

Table of Contents

Interface and Language Issues in Intelligent Systems for People with Disabilities*

Kathleen F. McCoy

Computer and Information Sciences Department and
Applied Science and Engineering Laboratories
University of Delaware/duPont Hospital for Children
Newark, DE 19716
U.S.A.
mccoy@cis.udel.edu

1 Introduction

The papers in this section describe a diverse set of applications of various Artificial Intelligence (AI) techniques. The overriding theme of the papers is that of interfaces to language/communication for people who have disabilities which make it difficult for them to communicate using spoken language, or interfaces that use spoken language or some other means (e.g., eye-tracking) as one kind of input for controlling an environment for people whose physical disability precludes them from physically manipulating their environment.

Several of the papers can be seen as falling into the area of Augmentative and Alternative Communication (AAC), and many use some processing methodologies from the AI area of Natural Language Processing (NLP). Some represent "mature" technologies that have been tested on actual users, while others involve the development of technologies which hold future promise.

In this paper I will attempt to give an overview of the area to which many of these papers can be fit – pointing out places on which the papers in this volume can be seen as focusing and where application of AI technologies might continue. Next an overview of NLP will be provided (again pointing out which aspects the papers in this volume have emphasized). Finally, other AI areas emphasized in these papers will be discussed.

2 Augmentative and Alternative Communication (AAC)

AAC is the field of study concerned with providing devices or techniques to augment the communicative ability of a person whose disability makes it difficult to speak in an understandable fashion.

A variety of AAC devices and techniques exist today. Many of these are aimed at people who have severe speech impairments (such that their speech cannot

---

* This work has been supported by a Rehabilitation Engineering Research Center grant from the National Institute on Disability and Rehabilitation Research (#H133E30010). Additional support was provided by the Nemours Foundation.

V. O. Mittal et al. (Eds.): Assistive Technology and AI, LNAI 1458, pp. 1–11, 1998.
© Springer-Verlag Berlin Heidelberg 1998

be reliably understood) and whose muscular control makes typing on a standard keyboard difficult (if not impossible). Some devices designed for such populations are non-electronic word boards containing words and phrases in standard orthography and/or iconic representations. A user of such a non-electronic system points to locations on the board and depends on the listener to appropriately interpret the selection. Electronic word boards may use the same sorts of selectable items, but may also include speech synthesis. These presumably provide more independence for the user who does not need to rely on a partner to interpret the selections. However, these systems may place more burden on the user who must be aware of the actual strings associated with each selection and must ensure that the synthesized string be an appropriate English sentence. Since the system will only "speak" what has been selected, more selections are generally required per sentence and speed of selection becomes more crucial.

3   Computer-Based Augmentative and Alternative Communication

A traditional computer-based AAC system can be viewed as providing the user with a "virtual keyboard" that enables the user to select items to be output to a speech synthesizer or other application. A virtual keyboard can be thought of as consisting of three components: (1) a physical interface providing the method for activating the keyboard (and thus selecting its elements), (2) a language set containing the elements that may be selected, and (3) a processing method that creates some output depending on the selected items. All three of these elements must be tailored to an individual depending on his/her physical and cognitive circumstances and the task they are intending to perform.

For example, for people with severe physical limitations, access to the device might be limited to a single switch. A physical interface that might be appropriate in this case involves row-column scanning of the language set that is arranged (perhaps in a hierarchical fashion) as a matrix on the display. The user would make selections by appropriately hitting the switch when a visual cursor crosses the desired items. In row-column scanning the cursor first highlights each row moving down the screen at a rate appropriate for the user. When the cursor comes to the row containing the desired item, the user hits the switch causing the cursor to advance across the selected row, highlighting each item in turn. The user hits the switch again when the highlighting reaches the desired item in order to select it. For users with less severe physical disabilities, a physical interface using a keyboard may be appropriate. The size of the keys on the board and their activation method may need to be tailored to the abilities of the particular user.

One of the papers in this volume, [11], involves an intelligent eye-tracking system which can be viewed as a physical interface to an AAC system. The system allows a user to control a computer through five electrodes placed on the head. Users can be taught to control their muscles and to use head movement to control a cursor on the screen. The use of this EagleEyes system with appropriate applications (e.g., language sets) has enabled several users to communicate and exhibit intelligence that was previously locked out because their disabilities precluded their use of other traditional interfaces.

Great challenges in this work include (1) use of appropriate sensors, (2) developing methods for determining when eye gaze is being used purposefully (i.e., dealing with the "midas touch"), (3) accuracy and control, (4) developing augmentations such as mouse clicks via eye-blinks. Various training methods for using the interface are discussed and various applications developed and tailored to individuals.

The physical interface is also somewhat of an issue in [27] also in this volume. This paper focuses on a modern AAC device which uses vision techniques to recognize sign language. The eventual application is that of translating the signed material into spoken text, allowing the person who is signing to be understood by people who do not know sign language.

The interface issue in this paper is that the authors envision the recognition system to be a "wearable computer" which is worn by the signer. The camera for the system described in the paper is worn on a cap and has a view of the signer's hands (which are tracked by the vision system). While the authors note that the eventual system will need to capture facial expression, the cap mounted system has shown greater accuracy in picking out and following the hands than their previous attempts. This has led to greater overall system accuracy.

While not an AAC system, [15] is an excellent example of a system whose interface combines several modalities (e.g., spoken and gestural) in allowing the user to control a robot to manipulate the environment.

Independent of the physical interface in an AAC system is the language set that must also be tuned to the individual. For instance, the language set might contain letters, words, phrases, icons, pictures, etc. If, for example, pictures are selected, the processing method might translate a sequence of picture selections into a word or phrase that will be output as the result of the series of activations. Alternatively, consider a language set consisting of letters. A processing method called abbreviation expansion could take a sequence of key presses (e.g., chpt) and expand that set into a word (e.g., chapter).

The use of a computer-based AAC device generally has many trade-offs. Assuming a physical interface of row-column scanning, a language set consisting of letters would give the user the most flexibility, but would cause standard message construction to be very time consuming. On the other hand, a language set consisting of words or phrases might be more desirable from the standpoint of speed, but then the size of the language set would be much larger causing the user to take longer (on average) to access an individual member. In addition, if words or phrases are used, typically the words would have to be arranged in some hierarchical fashion, and thus there would be a cognitive/physical/visual load involved in remembering and accessing the individual words and phrases.

One kind of language set that has been found to be very effective is an iconic language set. An iconic language set must be coupled with a processing method to translate the icon sequence selected into its corresponding word/phrase. A challenge in developing an iconic language set is to develop a language that can be easily used. In particular, the user must be able to recall the sequences of icons that produce the desired output. [1] is a paper in this volume concerned with a design methodology for developing iconic languages. In the methodology, icons in the language are associated with a set of semantic features which capture the various semantic concepts inherent in an icon. A set of relations is described which allow the meanings of individual icons to be combined in various fashions. The kinds of combinations available and the resulting semantic inferences can be used to establish meaningful sequences of icons and to predict the resulting intuitive meaning.

[31] (this volume) is concerned with several aspects of an AAC system that must be tuned if the language set consists of phrases rather than individual lexical items. In particular, when phrases/sentences are used the number of items to be accessed is quite large and the time spent navigating to the phrase must be minimal. This is because if the phrase takes longer to access than it would have taken to compose it from scratch, there is no savings!

The key idea in [31] is to store the text needed for a typical event (e.g., going to a restaurant) with the typical sub-events for which the text might be needed. For example, if a typical restaurant script has an entering, ordering, eating, and leaving scene, the text needed for each of those scenes would be stored with the scene. Thus the user could access the appropriate text by following along the script. Such a system puts certain requirements on the system interface, and some of these are explored in the paper (as well as a preliminary evaluation of the use of schemata to store prestored text in a communication aid).

The paper [20] is focused on the processing aspect of an AAC system. One issue that must be faced concerns literacy skills for people who use AAC systems. Because of the enormous time required to communicate with an AAC device, many users develop strategies for getting across their functional communication using telegraphic utterances. While this is a very beneficial strategy, it may cause non-standard English to be reinforced. The idea in this paper is to use processing on the telegraphic selections given by the user in order to give correct English sentence feedback. Such a system may have the benefit of raising the literacy skills of the user. The expansion of telegraphic input into full English sentences has been discussed in previous papers by this group. The focus of [20] is on additions (such as a user model which captures levels of literacy acquisition) which would be necessary for this new application.

While not a traditional AAC system, [28] focuses on the processing required to translate Japanese into Japanese Sign Language (JSL) so as to make Japanese communication accessible to a person who is deaf and cannot understand Japanese. The basic methodology in translating between the two languages involves a translation of lexical items (word-for-word translation) and a translation of syntactic structures between the two languages. One problem that is of concern is that there may not be a lexical item in JSL corresponding to a particular lexical item in Japanese. The paper describes a method for finding a similar word based on some meaning information contained in the dictionaries for Japanese and for Japanese Signs contained in the system. They have evaluated their system on some news stories and are reaching translation accuracy rates of 70%.

Another paper whose aim is similar to [28] is [6] which is concerned with knowledge bases necessary for a vision system to translate American Sign Language into English. In particular, the focus of the paper is at the word level and it is concerned with capturing information which would allow the signs to be translated into their corresponding English word equivalents. This is done using a feature-based lexicon which captures linguistically motivated features of the signs (which may be recognized by the vision system). This allows the vision system to search for the word corresponding to a sign in an efficient manner.

4   The Application of NLP

The fields of Natural Language Processing (NLP) and Computational Linguistics attempt to capture regularities in natural (i.e., human) languages in an effort to enable a machine to communicate effectively with a human conversational partner [2,3,12,13]. Areas of research within NLP have concentrated on all levels of processing – from the sub-word level (e.g., phonology, morphology) all the way up to the discourse level.

Chapman [6], in this volume, takes advantage of linguistic work concerning individual signs and their components in American Sign Language (ASL). The goal is to develop a sign language lexicon which can be used to recognize ASL signs (and to translate them into their English equivalents). The lexicon would be used by a vision system, and it indexes the signs by both manual and non-manual information. The manual information includes information about the movement, location (with respect to the signer's body), handshape, and hand orientation used in making the sign. Non-manual information includes facial characteristics (such as raised eyebrows) or body orientation during the sign. These choices were motivated by sign formation constraints.

Above the word level, three major areas of research in NLP and Computational Linguistics (syntax, semantics, and pragmatics) deal with regularities of language at different levels. Various techniques have been developed within each which will be useful for application to various AAC technologies.

4.1   Syntax

The syntax of a language captures how the words can be put together in order to form sentences that "look correct in the language" [2]. Syntax is intended to capture structural constraints imposed by language which are independent of meaning. For example, it is the syntax of the language that makes:

"I just spurred a couple of gurpy fliffs."

seem like a reasonable sentence even if some words in the sentence are unknown, but makes

"Spurred fliff I couple a gurpy."

seem ill-formed.

Processing the syntax of a language generally involves two components: 1) a grammar which is a set of rules that refer to word categories (e.g., noun, verb) and various morphological endings (e.g., +S for plural, +ING) that capture the allowable syntactic strings in a language and; 2) a parser which is a program that, given a grammar and a string of words, determines whether the string of words adheres to the grammar. (See [2,3,12,37] for examples of various parsing formalisms and grammars.)

Using a grammar and parser an AAC system would be able to: 1) determine whether or not the utterance selected by the user was well-formed syntactically, 2) determine valid sequences of word categories that could form a well-formed sentence, 3) given a partial sentence typed by the user, determine what categories of words could follow as valid sentence completions, 4) determine appropriate morphological endings on words (e.g., that a verb following the helping-verb "have" must be in its past participle form), and 5) determine appropriate placement of function words which must be added for syntactic reasons (e.g., that certain nouns must be preceded by an article, that the actor in a passive sentence is preceded by the word "by").

Syntactic knowledge is currently being successfully applied in a number of AAC projects. For example, several word prediction systems use syntactic information to limit the words predicted to those which could follow the words given so far in a syntactically valid sentence [22,30,29]. To some extent, many grammar checkers available today and systems aimed toward language tutoring (e.g., [24,25,19,33]) also use syntactic information, though there is still great room for improvement.

In this volume syntactic processing of spoken language is used in [15] in order to understand the user's intentions. In that system, a side-effect of parsing is the computation of meaning. Following a grammar of sign language sentences (in this case, there is only one sentence pattern) is used in [27] in order to aid the Hidden Markov Model to recognize the signs. Finally, [28] use syntactic translation rules as one step in translating Japanese sentences into Japanese Sign Language sentences.

4.2   Semantics

The area of semantics deals with the regularity of language which comes from the meanings of individual words and how the individual words in a sentence form a meaningful whole. A problem in semantics is the fact that many words in English have several meanings (e.g., "bank" may refer to the edge of a river or to a financial institution). In Computational Linguistics the use of selectional restrictions [16], case frames [9,10], and preference semantics [36] is based on the idea that the meanings of the words in a sentence are mutually constraining and predictive [26]. When the words of a sentence are taken as a whole, the meanings of the individual words can become clear.

Consider the sentence "John put money in the bank." Here the financial institution meaning of "bank" can be inferred from the meaning of the verb "put" (which expects a thing to be put and a location to put it in) and the fact that "money" is the appropriate kind of object to be put in a financial institution.

Note that in order to take advantage of semantics, a natural language processing system must (1) have rules (selectional restrictions, case frames) which capture the expectations from individual words (e.g., "eat" is a verb that generally requires an animate agent and an object which can be classified as a food-item), and (2) have a knowledge base that contains concepts that are classified according to their meanings (e.g., "apples" are food-items, "John" is a person, and "people" are animate).

The Compansion system [8,18], which is the underlying system referred to in [20], has made extensive use of semantic information to transform telegraphic input into full sentences. In this volume it is suggested that the full sentence constructed might be used as a literacy aid. Semantic information is also a main component of the PROSE [34] system developed at the University of Dundee. PROSE is intended to give the user access to prestored phrases/sentences/stories which can be accessed according to their semantic content. The basic idea is that sets of phrases, stories, sentences etc. will be input by the user (in advance) along with some semantic information about their content. PROSE will then store this information in an intelligent way according to the semantic information given. The system will then retrieve the pre-stored material, based on minimal prompting by the user, in semantically appropriate contexts.

Both syntax and semantic information are used in the project described in [7,5] involving "co-generation" (where the generation of a natural language sentence is shared between the user of the system and the system itself). This project attempts to speed communication rate by allowing sentences to be generated with fewer selections on the part of the user. Here the user fills in a "semantic template" with desired content words. The system then generates a full grammatical sentence based on the semantic information specified by the user.

In this volume two papers make extensive use of semantic information for diverse purposes. [1] uses semantic information associated with icons and combination methods to determine "natural meanings" in sequences of icons. It is suggested that the combination rules can be useful in developing intuitive iconic languages.

In describing a machine translation system between Japanese and Japanese Sign Language (JSL) [28] uses semantic information in order to find an appropriate translation for a Japanese word when there is no corresponding word in JSL. In order to do this, they look for similar Japanese words (i.e., those with the same concept identifier) in an extensive Japanese dictionary and attempt to find a word that does have a translation in JSL. Failing this, they attempt using words in the dictionary definition of the word with no JSL equivalent. Finally, if this fails as well, they attempt to use the superconcept. If all of these methods fail, the system resorts to using finger-spelling. The finding of an appropriate substitution word is possible because of the semantic information encoded in their Japanese dictionary.

4.3 Pragmatics

Pragmatic information refers to the broad context in which language and communication takes place [2,14,17]. Situational context and previous exchanges produce conversational expectations about what is to come next. Natural language processing has concerned itself with developing computational mechanisms for capturing these same expectations in a computer.

A great deal of AAC work that takes advantage of pragmatic information has come from the University of Dundee. Their CHAT system [4] is a communication system that models typical conversational patterns. For example, a conversation generally has an opening consisting of some standardized greeting, a middle, and a standardized closing. The system gives users access to standard openings and closings (at appropriate times). In addition (for the middle portion of a conversation) it provides a number of "checking" or "fill" phrases (e.g., "OK", "yes") which are relatively content free but allow the user to participate more fully in the conversation.

The TALKSBACK system [32,35,34] incorporates user modeling issues. It takes as input some parameters of the situation (e.g., the conversational partners, topics, social situation) and predicts (pre-stored) utterances the user is likely to want based on the input parameters. For example, if the user indicates a desire to ask a question about school to a particular classmate, the system might suggest a question such as "What did you think of the geography lesson yesterday?". In other words, the system attempts to use the parameters input by the user to select utterances that are pragmatically appropriate.

Pragmatic information is the key in [31] in this volume. In their system pre-stored text is stored in schema structures [23,21] which capture typical sequences of events. This should allow access to text appropriate for an event by allowing the system user to "follow along" the typical sequence.

Pragmatic information in the form of a user model is also a focus of [20]. Here the user model attempts to capture the level of literacy acquisition in an attempt to provide beneficial feedback to the user.

5 Other Artificial Intelligence Technology

Another AI technology prominent in this section is vision processing. [27] uses vision technology in order to identify and track the hands on a video, and interprets sign language using Hidden Markov Models (HMM's). Evaluation of the results includes an experiment where gloves are worn (and the hand is tracked by color) and an experiment where the hands are tracked on skin tone. In both experiments the HMM is trained on 400 sentences and 100 sentences are used for testing. The training and testing sets use a limited number of signs, and the recognition rate is above 97% when the sentence structures are constrained to follow a predefined sentence pattern.

The paper [15] focuses on the integration of several different AI technologies in order to provide an interface that enables a person who has physical disabilities to manipulate an unstructured environment. The project combines a vision subsystem (which is able to identify object location, shape, and pose), an interface subsystem which interprets limited spoken commands combined with pointing gestures from a head-mounted pointing device, and a planning subsystem that interprets commands by the user and plans a method for carrying out the request. The user of the system provides some information to the system (such as indicating the class of particular objects) and then can ask the system to move objects around to various locations.

6 Conclusion

The papers in this volume provide us with a snapshot of the variety of issues that must be considered when applying AI technologies to projects involving people with disabilities. Here a focus is on controlling interfaces and language issues. The AI technologies used are quite varied; the resulting ideas have a great deal of promise and point us to future possibilities.

References

1. P. L. Albacete, S. K. Chang, and G. Polese. Iconic language design for people with significant speech and multiple impairments. In *This Volume*. 1998.
2. J. Allen. *Natural Language Understanding*. Benjamin/Cummings, CA, 1987.
3. J. Allen. *Natural Language Understanding, Second Edition*. Benjamin/Cummings, CA, 1995.
4. N. Alm, A. Newell, and J. Arnott. A communication aid which models conversational patterns. In Richard Steele and William Gerrey, editors, *Proceedings of the Tenth Annual Conference on Rehabilitation Technology*, pages 127–129, Washington, DC, June 1987. RESNA.
5. A. Copestake, D. Flickinger, and I. Sag. Augmented and alternative NLP techniques for augmentative and alternative communication. In *Proceedings of the ACL Workshop on Natural Language Processing for Communication Aids*, Madrid, Spain, 1997.
6. R. N. Chapman. Lexicon for ASL computer translation. In *This Volume*. 1998.
7. A. Copestake. Applying natural language processing techniques to speech prostheses. In *Working Notes of the 1996 AAAI Fall Symposium on Developing Assistive Technology for People with Disabilities*, Cambridge, MA, 1996.
8. P. W. Demasco and K. F. McCoy. Generating text from compressed input: An intelligent interface for people with severe motor impairments. *Communications of the ACM*, 35(5):68–78, May 1992.

9. C. J. Fillmore. The case for case. In E. Bach and R. Harms, editors, *Universals in Linguistic Theory*, pages 1–90, New York, 1968. Holt, Rinehart, and Winston.

10. C. J. Fillmore. The case for case reopened. In P. Cole and J. M. Sadock, editors, *Syntax and Semantics VIII: Grammatical Relations*, pages 59–81, New York, 1977. Academic Press.

11. J. Gips. On building intelligence into EagleEyes. In *This Volume*. 1998.

12. G. Gazdar and C. Mellish. *Natural language processing in lisp, an introduction to computational linguistics*. Addison-Wesley Publishing Company, New York, 1989.

13. R. Grishman. *Computational linguistics: An introduction*. Cambridge University Press, Cambridge, 1986.

14. A. Joshi, B. Webber, and I. Sag, editors. *Elements of discourse understanding*. Cambridge University Press, Cambridge, 1981.

15. Z. Kazi, S. Chen, M. Beitler, D. Chester, and R. Foulds. Speech and gesture mediated intelligent teloperation. In *This Volume*. 1998.

16. J. J. Katz and J. A. Fodor. The structure of semantic theory. *Language*, 39, 1963.

17. S. Levinson. *Pragmatics*. Cambridge University Press, Cambridge, 1983.

18. K.F. McCoy, P.W. Demasco, M.A. Jones, C.A. Pennington, P.B. Vanderheyden, and W.M. Zickus. A communication tool for people with disabilities: Lexical semantics for filling in the pieces. In *Proceedings of the First Annual ACM Conference on Assistive Technologies (ASSETS94)*, pages 107–114, Marina del Ray, CA:, 1994.

19. C. Morris, A. Newell, L. Booth, and I. Ricketts. Syntax PAL: A writing aid for language-impaired users. In *Presented at ISAAC-92*, volume 8, Philadelphia, PA, 1992. ISAAC, Abstract in AAC – Augmentative and Alternative Communication.

20. Christopher A. Pennington and Kathleen F. McCoy. Providing intelligent language feedback for augmentative communication users. In *This Volume*. 1998.

21. R. C. Schank and R. P. Abelson. *Scripts, plans, goals and understanding: An inquiry into human knowledge structures*. Erlbaum, Hillsdale, NJ, 1977.

22. A. L. Swiffin, J. L. Arnott, and A. F. Newell. The use of syntax in a predictive communication aid for the physically impaired. In Richard Steele and William Gerrey, editors, *Proceedings of the Tenth Annual Conference on Rehabilitation Technology*, pages 124–126, Washington, DC, June 1987. RESNA.

23. R.C. Schank. *Dynamic Memory: A theory of reminding and learning in computers and people*. Cambridge University Press, NY, 1982.

24. L.Z. Suri and K.F. McCoy. Correcting discourse-level errors in CALL systems. In *Proceedings of The World Conference on Artificial Intelligence in Education (AIED-93) (refereed poster presentation)*, August 1993.

25. L.Z. Suri and K.F. McCoy. A methodology for developing an error taxonomy for a computer assisted language learning tool for second language learners. Technical Report TR-93-16, Dept. of CIS, University of Delaware, 1993.

26. S. Small and C. Rieger. Parsing and comprehending with word experts (a theory and its realization). In W.G. Lehnert and M.H. Ringle, editors, *Strategies for Natural Language Processing*, pages 89–147. Lawrence Erlbaum Associates, 1982.

27. T. Starner, J. Weaver, and A. Pentland. A wearable computer based American Sign Language recognizer. In *This Volume*. 1998.

28. M. Tokuda and M. Okumura. Towards automatic translation from Japanese into Japanese Sign Language. In *This Volume*. 1998.

29. J.A. VanDyke. Word prediction for disabled users: Applying natural language processing to enhance communication. Thesis for honors bachelor of arts in cognitive studies, University of Delaware, Newark, DE, 1991.

30. J. VanDyke, K. McCoy, and P. Demasco. Using syntactic knowledge for word prediction. Presented at ISAAC-92. Abstract appears in Augmentative and Alternative Communication, 8., 1992.

31. P.B. Vanderheyden and C.A. Pennington. An augmentative communication interface based on conversational schemata. In *This Volume*. 1998.

32. A. Waller, N. Alm, and A. Newell. Aided communication using semantically linked text modules. In J. J. Presperin, editor, *Proceedings of the 13th Annual RESNA Conference*, pages 177–178, Washington, D.C., June 1990. RESNA.

33. A. Wright, W. Beattie, L. Booth, W. Ricketts, and J. Arnott. An integrated predictive wordprocessing and spelling correction system. In Jessica Presperin, editor, *Proceedings of the RESNA International '92 Conference*, pages 369–370, Washington, DC, June 1992. RESNA.

34. A. Waller, L. Broumley, and A. Newell. Incorporating conversational narratives in an AAC device. Presented at ISAAC-92. Abstract appears in Augmentative and Alternative Communication, 8, 1992.

35. A. Waller, L. Broumley, A. F. Newell, and N. Alm. Predictive retrieval of conversational narratives in an augmentative communication system. In *Proceedings of 14th Annual Conference*, Kansas City, MI, June 1991. RESNA.

36. Y. Wilks. An intelligent analyzer and understander of English. *Communications of the ACM*, 18(5):264–274, 1975.

37. T. Winograd. *Language as a cognitive Process, Volume 1: Syntax*. Addison-Wesley Publishing Company, Reading, Ma, 1983.

Iconic Language Design for People with Significant Speech and Multiple Impairments

P. L. Albacete[1], S. K. Chang[1], and G. Polese[2]

[1] Department of Computer Science
University of Pittsburgh
Pittsburgh, PA 15260
U.S.A.
[2] Dipartimento di Informatica ed Applicazioni
Universitá di Salerno,
84081 Baronissi (SA)
Italy

Abstract. We present an approach of iconic language design for people with significant speech and multiple impairments (SSMI), based upon the theory of Icon Algebra and the theory of Conceptual Dependency (CD) to derive the semantics of iconic sentences. A knowledge based design environment supporting the phases of this approach is described.

1   Introduction

Iconic languages are a specific type of visual languages, they have been used successfully in human-computer interface, visual programming, and human-human communication. The iconic language used in human-computer communication usually has a limited vocabulary of visual icons and a specific application domain: database access, form manipulation, image processing, etc. There are also iconic languages for human-human communication used in *augmentative communication* by people with SSMI. Finally, there are "natural" iconic languages such as the Chinese ideographs, the Mayan glyphs and the Egyptian pictograms.

In [13] we presented a design methodology for iconic languages, and exercised it in the design of the iconic languages of the $Minspeak^{TM}$ systems for augmentative communication. The design methodology serves two purposes. First of all, it is a descriptive model for the design process of the iconic languages used in the $Minspeak^{TM}$ systems. Second, it is also a prescriptive model for the design of other iconic languages for human-machine interface. The experience learned from the design of the iconic languages for the $Minspeak^{TM}$ systems has led to some valuable insight in the design of visual languages in general. We could successfully extend its theoretical framework by adding temporal knowledge and a syntactic framework for visual sentences to accommodate the design of general visual languages [14]. We think that the extension of this framework to the design of visual languages for multimedia environments [12] will provide in future means to construct more powerful systems for augmentative communication.

V. O. Mittal et al. (Eds.): Assistive Technology and AI, LNAI 1458, pp. 12–32, 1998.
© Springer-Verlag Berlin Heidelberg 1998

There are a variety of augmentative communication systems for people with physical or cognitive limitations, ranging from unaided communication such as American Sign Language, to aided communication systems such as computerized voice output systems. The $Minspeak^{TM}$ systems conceived by Bruce Baker use the principle of *semantic compaction* [2,3,4,7]. It involves mapping concepts on to multi-meaning icon sentences and using these icon sentences to retrieve messages stored in the memory of a microcomputer. The stored messages can be words or word sequences. A built-in speech synthesizer can then be used to generate the voice output. Over the past ten years, more than 20,000 $Minspeak^{TM}$ units have been distributed all over the world. Swedish, German, Italian and other $Minspeak^{TM}$ systems have been developed.

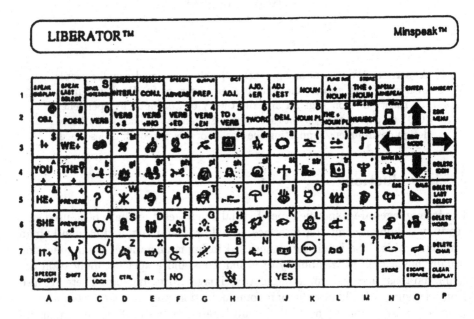

Fig. 1. A $Minspeak^{TM}$ Iconic Keyboard.

A $Minspeak^{TM}$ Iconic keyboard ($WordStrategy^{TM}$) is shown in Figure 1. When the user touches the icons on the keyboard, the system produces the voice output. Thus the $Minspeak^{TM}$ keyboard can serve as an augmentative communication system. For example, when the APPLE icon and the VERB icon are depressed in that order, the system produces the voice output "eat." The sequence "APPLE VERB" is called an *iconic sentence*. A different iconic sentence such as "APPLE NOUN" will produce the voice output "food." The APPLE icon by itself is thus ambiguous. The basic idea of *semantic compaction* is to use ambiguous icons to represent concepts. For example, the APPLE icon can represent "eat" or "food." Ambiguity is resolved when several icons are combined into an iconic sentence. This principle allows the representation of many concepts (usually around two thousand) using a few icons (usually around fifty). The concepts form what we call the domain language. Thus, our goal in the design of an iconic language is to provide a visual representation to the domain language, which entails representing each domain concept through an iconic sentence semantically close to it.

Our approach is to formalize the methodology to design iconic languages, based upon the theory of *Icon Algebra* and the theory of *Conceptual Dependency*. We will show how these theories have been exploited to derive the semantics of iconic sentences, which is a critical step in the design methodology mentioned above. The formalization of the design methodology has led to a deeper understanding of iconic languages and iconic communications, allowing to further explore theoretical issues in this framework.

The paper is organized as follows. Section 2 begins by explaining the frame-based method used in our knowledge representation. Then it introduces the theory of *Icon Algebra* [9,11] as a formal approach for deriving the meaning of iconic sentences. It ends with a brief description of the theory of *Conceptual Dependency* to serve as the semantic model of iconic sentences. The extension of CD forms for iconic languages is discussed in Section 3. The algorithm for making inferences using the icon algebra and the CD forms is presented in Section 4. In Section 5, we discuss the practical significance of this research and its implications to iconic language design for people with speech disabilities. Finally, in the appendix we provide some experimental results.

2   Background

Our approach for knowledge representation uses the frame-based method. The frame based method is very suitable, because it provides a unique way to represent knowledge of icons, iconic sentences, general visual sentences, concepts, and multimedia information. Thus, it provides a unified structure for representing all the different objects we will be handling in our methodology [13], which will simplify the definition of central functions such as the similarity function and inference algorithms for understanding visual sentences.

In the case of the Minspeak visual languages, the knowledge underlying an icon is represented using several metaphors, depending on the particular context of the visual language under development. According to these metaphors, one or more attributes are set in a frame structure which will be used to describe the semantics of an icon. We have made use of the metaphors from the *Types of Semantic Relationships* diagram [5], and have represented them through a frame structure. For instance, the metaphor "quantity" can be expressed through the two attributes "mass_quantity" and "count_quantity." If we want to express the knowledge in the icon "elephant," among the various attributes, we can fill the attribute "mass_quantity" with the value "big."

We use the frame-based structure for representing knowledge of words, icons or iconic sentences. In a broader context, we can also use frames for representing multimedia knowledge. Therefore, in what follows, we will use the term *object* to refer to any type of information from a given application domain. The frame of an object is filled with the values of the attributes describing the object, according to a set of metaphors, together with a *fuzzy parameter* in the range [0,1], denoted by Ω. Ω is the relevance of an attribute value in describing the given object. For instance, the value "Orange" of the attribute "Color" will be more relevant than the value "Round" of the attribute "Shape" for the object "Orange Fruit." The construction of the frame for each single object to be encoded by the visual language is done by the designer. The frame for each single icon is obtained by inheriting the frame of the object directly associated to the icon [8], and augmenting it with the additional meanings conveyed by the icon image. Then, we derive the frame for the iconic sentence by using an inference algorithm [13]. Examples are given in the appendix.

2.1 The Icon Algebra for Deriving Icon Semantics

Icon algebra [10] provides a powerful way to derive new meanings of an icon, or an iconic sentence, by applying some formal operators on it. An icon X is seen as a pair (X_m, X_i) where X_m represents the meaning of the icon, or the logical part, and X_i represents the image, or the physical part. An essential characteristic of an icon is that the logical part and the physical part are mutually dependent. That is, if the image is changed, its meaning will also be changed and vice versa. These concepts have been recently extended to model multimedia environments [12]. According to this reformulation of the theory, an icon can represent not only pictures, but also text, sound, and video. So for example, the physical part of an ear icon will be a sound, and different types of icons can be combined to form a multimedia presentation.

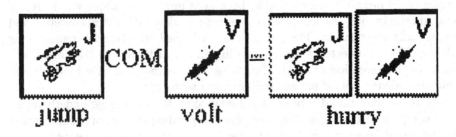

Fig. 2. The Iconic Operator COM.

The icon algebra applies some formal operators to icons to derive new icons. An example of the operator COM is shown in Figure 2. There, the two icons "jump" and "volt" both have the meaning "hurry" among their possible meanings, although this is not the primary meaning in either of them. Therefore only their combination leads to the concept "hurry." In the above, "jump" leads one to recall the concept "fast," "volt" has the quality "fast," and the resultant composite icon has the meaning "hurry."

We have formalized the semantics of a set of Icon Algebra operators, which can be combined with the primary meanings of the elementary icons composing an iconic sentence, to derive its possible meanings. The derived meanings are submitted to the designer who will decide whether to accept or discard the new meanings. The designer may also perform some actions such as assigning the Ω parameter or the appropriate slot that should contain the new information.

The semantic definition of the iconic operator COM is given below [13]. The operator COM may have the word attributes MERGED_WITH, AND, or WITH, because it semantically combines the meaning of two icons. For example, combining the icons "Jump" and "Volt" by applying COM to X.[is]jump.[recall]fast and Y.[is]volt.[quality]fast, we obtain the slot value fast_MERGED_WITH_fast which implicitly means "hurry." Again, the designer can make that derived meaning explicit by directly assigning it to the frame of the iconic sentence, or by performing a similarity inference. In the latter case, the similarity with the frame of the word 'hurry' should be detected.

We now explain the operators MAR, CON, ENH and INV. The marking operator MAR marks the image of the icon Y with the image of the icon X to emphasize a local feature. Here the first icon plays the role of "marker image." As an example, the Chinese character for "root" is an iconic sentence "tree" "low_marker," because Y.[is]tree.[part]root.[location]low marked by X.[is]low-marker.[recall]low results in Z.[is]root. Thus, marking is a *conceptual restriction* to extract an important local feature.

For the contextual interpretation operator CON, the meaning of the icon X is considered in the context of Y, and the result is usually a *conceptual refinement* (specialization) of the meaning of X. For example, given the iconic sentence "apple" "morning," since "apple" recalls the concept of "food," "food" in the "morning" leads to "breakfast." Therefore X.[is]apple.[is_a_concrete]food in the context of Y.[time]morning results in Z.[is]breakfast, and "breakfast" is a subclass of "food" in the hierarchy.

The enhancement operator ENH enhances the conceptual richness of the icon X by adding attributes from Y, and the result is usually an enrichment of the meaning of X. For the iconic sentence "thermometer" "thumb_down," since "low temperature" means "cold," X.[is]thermometer.[use]temperature enhanced by Y.[is]thumb_down.[recall]low leads to Z.[is]cold.

For the inversion operator INV, the meaning of the icon X is inverted. As the inversion is an unary operator, we use an icon image to represent the operator. For example, given the iconic sentence "knot" "god," since God stands for "true," the negation of "true" is "false." In the $Minspeak^{TM}$ iconic language, the icon "knot" stands for negation because its pronunciation the same sound as "not."

2.2 Conceptual Dependency

In this section we discuss those aspects of the theory of Conceptual Dependency that we will be using for Iconic Language processing. For a more detailed description of the theory see [16].

Conceptual Dependency (CD) is a theory of natural language and of natural language processing. It was created by Schank and can be employed for the construction of computer programs capable of understanding sentences of natural language, summarizing them, translating them into another language, and answering questions about them. The basic axiom of the theory is:

For any two sentences that are identical in meaning, regardless of the underlying language, there should be only one representation.

Thus, any information in the sentence that is implicit must be made explicit in the representation of the meaning for that sentence. In this theory, understanding of concepts is performed by mapping linear strings of words into conceptual structures. A conceptual structure is a kind of semantic net. It is defined as a network of concepts, where some classes of concepts may have specific relationships with other classes of concepts. This conceptual structure is called a conceptualization or CD form. A conceptualization can be active or stative. An active conceptualization consists of the following slots: actor, action, object, and direction. The latter is subdivided into source and destination. A stative conceptualization consists of the following slots: object, state, and value.

The slots can be filled by the following conceptual categories, see [16] for more details:

PPs: Picture Producers. Only Physical objects are PPs. They may serve in the role of objects as well as source, destination and recipient of actions.

ACTs: Actions. Actions can be done by an actor to an object.

LOCs: Locations. They are considered to be coordinates in space and can modify conceptualizations as well as serve as sources and destinations.

Ts: TIMES. The time is considered to be a point or a segment on a time line.

AAs: Action Aiders. AAs are modifications of features of an action.

PAs: Picture Aides, or Attributes of an Object. A PA is an attribute characteristic such as color or size plus a value for that characteristic. PAs can be used for describing a Physical object.

The ACTs category is composed by a set of primitive actions onto which verbs in a sentence can be mapped. This greately simplifies the way in which the rest of the slots in the CD forms can be filled. A complete list of actions and their descriptions are given in [16,17].

Rule 1. Certain PPs Can ACT. For example, the sentence "Kevin walked" may be represented using the primitive act PTRANS as

Actor: Kevin
Action: PTRANS
Object: Kevin
Direction:
           From: Unknown To: Unknown

Rule 2. PPs and Some Conceptualizations Can Be Described By an Attribute . For example, the sentence "Nancy is heavy" may be represented using the following stative conceptualization:

> Object: Nancy
> State: WEIGHT
> Value: Above the Average

One of the most important problems in natural language processing is inference. Schank used the primitive ACTs to organize and reduce the complexity of the inference process. He concerned himself with two kinds of inferences: results from ACTs and enablements for ACTs. Rules may be attached to be activated whenever the primitive is encountered. For example, given the textual sentence "Paul hit John with a rock," a mapping onto a CD form in which the Action slot is filled by the primitive action PROPEL, will trigger a rule that will allow the inference "John was hurt."

3   CD Forms for Iconic Languages

We have mentioned in the Background section that icon algebra provides a powerful way of deriving new meanings of an icon, or an iconic sentence, by applying some formal operators on it. The application of an operator to an iconic sentence will provide an intelligent selection of attributes from the frames of the single icons composing the sentence, which will be the basis for constructing the semantics of the whole sentence. For example, when applying the CON (context) operator to the icons "apple" and "rising-sun," the designer selects the frame attribute [is-a-concrete].food from the "apple" frame and the attribute [time].morning from the "rising-sun" frame, and applying the operator s/he can infer the meaning "food in the context of the morning" which is "breakfast." Another selection of frame attributes, such as [color].red from the "apple" frame and [location].horizon from the "rising-sun" frame, with the context operation will lead to "read in the context of the horizon" which may have no meaning.

This denotes the way we make use of Conceptual Dependency Theory (CDT). Conceptual Dependency Theory (CDT), as it was explained in the Background section, was conceived to enable a system to process and "understand" natural language. The basic way of allowing for understanding of a sentence is by providing a conceptual structure that can capture the meaning of a given textual sentence. This conceptual structure also facilitates the natural inferences that can be performed on the implicit meaning of the conceptualized textual sentence. The conceptual structure is described through CD forms. In our approach, we follow the same idea as in Conceptual Dependency Theory, but we use it for visual language processing and understanding. We process an iconic sentence, rather than a textual sentence. However, as it is done in CDT, to understand the iconic sentence we provide it with a conceptual structure, that captures the semantics of the sentence. The conceptual structure is described through a modified version of CD forms. While in CDT the CD forms to be associated to a textual sentence where chosen according to the verbs contained in the sentence, in our case they are chosen according to the Icon Algebra operators used to semantically compose the elementary icons in the iconic sentence. Moreover, the inference routines of traditional CD forms would modify the properties (attributes) of the associated textual sentence to produce the logical consequences implied by the semantics of the sentence. On the other hand, the inference routines of CD forms associated to iconic operators select and process a proper subset of attributes from the frames of the elementary icons in the iconic sentence, in order to produce its possible meanings.

Since one of the central aspects of our methodology is to capture the meanings of an iconic sentence, we will first discuss some of the features that are characteristic of the semantics of iconic sentences. These features are not necessarily shared by the semantics of textual sentences.

An important feature of an iconic sentence is that its meaning is inherently ambiguous, that is, its interpretation is not unique. This ambiguity arises from the fact that the icons composing the sentence can have multiple meanings and that they can be combined using all the different Icon Algebra operators. In fact, the operator suggests a way by which meanings of the icons composing the sentence should be combined to derive the meanings of the whole sentence. Thus, we decided to build our CD structures around the meaning of each operator.

In what follows, we describe how our CD forms were created to capture the possible meanings of an iconic sentence. In our approach, we have primitive connectors instead of primitive actions. Each operator has one or more corresponding connector/s, in a similar way as each verb in natural language has one or more corresponding primitive ACTs. The other slots in the CD forms vary according to the connector. However, as with the "actor" and "object" slots used in the CD forms of CDT, we identify two slots which are common to all of the CD forms. Basically, here we have two objects which are composed to form a third object according to the semantics of the connector in the CD form. These objects represent possible meanings of the icons in the sentence they represent. The first object can be viewed as an actor in the sense that it is the one which leads the inference process, whereas the second object plays the role of a passive modifier of the meanings derivable by the first object. In fact, the semantics of the sentence changes when we modify the order of the icons in the sentence. Thus, we could call these slots in order the "Active" and the "Passive" object. Then, the specific connector used might require to address more slots (as in the case of mark_has). In addition, each connector has a unique inference routine attached to it. What this routine does is to give a textual explanation of the possible meanings of the conceptualized visual sentence according to that connector.

As in CDT, we identified some conceptual categories. These categories define the possible fillers of the CD form slots. In our case, the category CONN is the only legal filler of the connector slot, and all the other categories can eventually fill the other slots. The rules that determine the possible combinations of the conceptual categories are defined by the semantics of each connector. The conceptual categories are:

CONN: connector. Establishes the relation between the active object and the passive object.

OBJ: object. Describes physical objects.

ENV: environment. Describes a context given by a time or a location.

MOD: modifier. Describes something that qualifies something else.

QUAL: quality. Defines a characteristic of something.

CAT: category. Defines a class of objects.

ACT: action. Defines something that is done.

The use of primitive connectors limits the number of interpretations of an iconic sentence. A list of the primitive connectors along with their corresponding Icon Algebra operators, follows:

IN_CONTEXT describes an object in the context of an environment. It corresponds to the iconic operator CON.

ENHANCED describes a quality enhanced by a modifier. It corresponds to the operator ENH.

INVERSION describes the negation of a quality. It corresponds to the operator INV.

MARK_IS_A describes an object that belongs to a certain category. It corresponds to the operator MAR.

MARK_HAS describes an object that has a feature that belongs to a given category. It corresponds to the operator MAR.

COM_LING_CONV describes something that is defined by the combination of a particular quality, namely a linguistic convention, with another word. It corresponds to the operator COM.

COM_REPEAT describes a concept by the repetition of a quality. It corresponds to the operator COM.

In CDT, each CD form has associated semantic constraints on the types of entities that can fill its slots. In a similar way, in our approach, there are semantic constraints on the type of entities that can fill the slots of a CD form, which are given by the semantics of the primitive connector associated to it. This connector determines not only the conceptual categories that can fill the CD form slots, but also the subset of icon's frame slots that will belong to the conceptual category. That is to say, that the semantics of the primitive connector is what determines which meaning/s of each icon in the iconic sentence is adequate for that CD form. It should be noted that if a different interpretation is given to the slots in the frame, then their correspondence to a given category may change.

Below, we describe each created CD form with the description of each slot and its possible fillers, and the corresponding inference routine.

1. CONNECTOR in_context
ACTIVE_OBJECT   OBJ   ?object
                 (OR Icon1[recall]
                     Icon1[is_a_concrete]
                     Icon1[use])
PASSIVE_OBJECT   ENV ?environment
                 (OR Icon2[location]
                     Icon2[time_seq])
INFERENCE   "?object in the ?environment"
                  "?object in the context of the ?environment"

As an example, given the iconic sentence "apple rising_sun," one of the possible inferences from this CD form will be "food in the context of the morning," where ?object=apple.[recall]food and ?environment=rising_sun.[time_seq] morning. This may be a way of expressing the concept "breakfast."

2. CONNECTOR enhanced
ACTIVE_OBJECT   QUAL   ? quality
                   (OR Icon1[recall]
                      Icon1[use])
PASSIVE_OBJECT   MOD   ? modifier
                   (OR Icon2[recall]
                      Icon2[mass]
                      Icon2[count])
INFERENCE   "?quality enhanced by ?modifier"

3. CONNECTOR inversion
ACTIVE_OBJECT   QUAL   ? quality
                   (OR Icon1[location]
                      Icon1[recall]
                      Icon1[cultural_conv]
                      Icon1[mass]
                      Icon1[is_a_concrete]
                      Icon1[is_an_abstract])
INFERENCE   "not ?quality"

Note: this primitive connector is only used when the second icon is the KNOT which is homonymous to "not," so the negation of the quality in the first icon is the meaning of the iconic sentence.

4. CONNECTOR mark_is

ACTIVE_OBJECT	CAT	? category
		(OR Icon1[recall]
		Icon1[is_a_concrete]
		Icon1[is_an_abstract]
		Icon1[cultural_conv])
PASSIVE_OBJECT	OBJ	? object
		(is_a ?category Icon2[?])
INFERENCE		"?category marks ?object"

Note: the expression (is_a ?category Icon2[?]) means that the selected value for the slot passive_object will be one that belongs to the category defined by the filler of the active_object slot.

5. CONNECTOR mark_has

ACTIVE_OBJECT	CAT	? category
		(OR Icon1[recall]
		Icon1[is_a_concrete]
		Icon1[is_an_abstract]
		Icon1[cultural_conv])
PASSIVE_OBJECT	QUAL	?quality1 Icon2[?]
PASS_OBJECT_HAS	QUAL	?quality2
		(is_a ?category ?quality1[?])
INFERENCE		"?quality1 has ?quality2 that belongs to ?category"

Note: the expression (is_a ?category ?quality1[?]) says that the filler of the slot pass_object_has will be a quality that is possessed by the filler of passive_object, and that belongs to ?category. For example, in the iconic sentence "rainbow chest," one of the meanings this CD form will capture is the following: ?category=rainbow.[recall]color ?quality1= chest.[quality]treasure ?quality2=treasure[color].gold, because the color gold is a quality possessed by treasure, and it belongs to the category color.

6. CONNECTOR com_ling_conv

ACTIVE_OBJECT	QUAL	?linguistic_conv
		Icon1[linguistic_conv]
PASSIVE_OBJECT	QUAL	?quality Icon2[?]
INFERENCE		"?linguistic_conv ?quality"

7. CONNECTOR com_repeat

ACTIVE_OBJECT	QUAL	?quality1 Icon1[recall]
PASSIVE_OBJECT	QUAL	?quality2
		(synonym ?quality1
		Icon2[?])
INFERENCE		"?quality1 and ?quality2"

Note: the expression (synonym Icon1[recall] Icon2[?]) means that the slot value of Icon2 should be a synonym of the selected slot value of Icon1.

4   An Inference Algorithm

In this section we describe the inference engine used within the design methodology for iconic languages [13], to derive the meanings of iconic sentences. This is an important step of the methodology, because it allows to search for iconic sentences closest in meaning to the objects of the domain language to be visualized. The engine basis its inferences on the CD forms as described in the former section. They basically provide a conceptual representation for iconic sentences, with attached inference routines to reflect the semantics of the iconic operator they are associated with.

The inference algorithm has three main modules. A basic module builds a skeleton frame for the iconic sentence by searching for common attribute values in the frames of the single icons in the sentence, according to three match criteria. Each criteria will have a different weight in a formula for computing the Ω values for the matching attributes [13], which will be stored on the skeleton frame for the iconic sentence.

A second module of the inference algorithm tries to extend the skeleton frame for the iconic sentence by applying iconic operators. This is done by running the CD forms associated to each operator. As seen in the previous section, a CD form for an operator contains information on how to build textual descriptions of the iconic sentence to which the operator is applied. Each description is given an Ω value taken as the minimum of the Ω values associated to the slot values used by the CD form to build the description. Besides, we also keep the operator name together with each explanation so as to be able to provide a formal rationale, based on icon algebra, to be stored into a dictionary for the final iconic language. The design environment let the designer edit these descriptions and assign them as slot values of the frame for the iconic sentence. We could also give them in input to a more powerful or specialistic inference mechanism to perform further semantic inferences on the iconic sentence.

A third module runs a brute force algorithm to provide further textual descriptions for the iconic sentence, aiming to furtherly enrich its semantic frame. The algorithm scans all the semantic attributes of the frames for the single icons in the iconic sentence, and applies semantic operators of icon algebra to build the new textual descriptions. The procedure for building these descriptions is similar to the one used by the second inference module, but this time the selection of the frame slots from the elementary icons in the sentence is no more driven by a CD form. Thus, the algorithm will operate on all the possible combinations of slots. If the designer is not satisfied with the textual descriptions generated with CD forms in the second inference module, the designer can choose to not run the module. S/he is allowed to interact with the system to provide cutting thresholds, such as the maximum number of new values that s/he wants the system to generate, and the lower bound on the parameter Ω for an attribute to be taken into consideration.

According to the CD forms associated to an operator, it makes sense to select only a restricted set of slot values to derive new textual descriptions. This somehow reflects the logic used in CD theory, where the semantics of a primitive action was expressed through inference routines, which would manipulate the values of the slots affected by that action. Basically, our CD forms indicate the slots which are more likely to contain the semantic information needed for the application of the given operator. As seen in the previous section, the number of these slots is very small as opposed to the number of slots processed by the brute force algorithm of the third inference module described above.

For iconic sentences of length greater than 2, the inference will be iterative, and more than one iconic operator might be applied between pairs of icons in the sentence. Obviously, the order chosen to process iconic operators and icons will affect the final result. We can provide a syntactic mechanism, like positional or relation grammars [15,18], for parsing the iconic sentences and their operators. The spatial relations between icons in the iconic sentence can represent explicit iconic operators. Thus, each of these spatial relations will have attached CD forms which will be triggered whenever the parser recognizes them.

The inference engine has been developed in Visual C++ language and runs on IBM compatible PC. It is part of a set of tools supporting the methodology for iconic language design [13]. The whole system takes in input a language (the domain language) to be encoded through iconic sentences, and a frame-based knowledge base associated to the domain language. In the case of $Minspeak^{TM}$ the domain language is a subset of a natural language. The system starts clustering the domain language according to a similarity function. The latter computes distances between objects (words, icons, iconic sentences, multimedia objects) by comparing their frames, and it can be customized according to the context. Therefore, the designer is provided with tools to design icons and their corresponding frames. At this point, s/he can run the inference engine to derive frames for the iconic sentences, which will be used to compute their conceptual similarity respect with objects of the domain language, in order to decide an appropriate encoding of domain objects through iconic sentences.

In the Appendix we show results of semantic inferences on some iconic sentences.

5   Discussion

The use of visual languages as augmentative communication tools for people with communication impairments has always been of interest. Back in the 1930s, C. K. Bliss [6] designed an international communication system based on a set of symbols, which was later adapted for non-vocal physically handicapped persons. Although there are several thousand Bliss symbols, a functional Bliss chart might contain several hundred symbols which are pointed to with a finger or a head pointer. Words that are not represented directly by a symbol are obtained by combining special symbols that indicate operations such as "similar sound" or "opposite-to," with other symbols. However, the listener must apply its own world knowledge to infer the intended word. This puts a great burden on the listener and may make a dialogue with a stranger ineffective.

The advent of computers facilitated the development of many types of augmentative communication systems. However, systems using abbreviations may become arbitrary when the vocabulary grows large, and systems using word prediction techniques require user's interactive selection. The $Minspeak^{TM}$ system does not rely on the listener's world knowledge, as the Bliss system does, to interpret the words that the user is trying to convey. It also does not use abbreviation or word prediction to encode the vocabulary. The success of this approach indicates that augmentative communication systems based upon visual languages deserve serious consideration. Moreover, with the recent extension of visual languages to handle multimedia information [12], we can think of extending our approach to handle not only textual domain languages, but also more general multimedia domain languages, leading to a broader class of applications.

There are three important aspects that determine the usefulness of this kind of systems:

1. expressive power of the system,

2. how easy it is to learn how to use the system, and

3. how easy it is to use the system.

The expressive power of the system refers to how much and how well a user can express his/her ideas through the system. The $Minspeak^{TM}$ system provides a predefined vocabulary consisting of several thousand words (in our every day conversations we use an average of one thousand words), and words not included in the vocabulary can be spelled out. However, the system only uses between 50 to 120 icons. This shows how powerful visual languages can be when they allow ambiguous icons. However, the level of satisfaction of a user of the system with regard to its expressive power is subjective and requires careful consideration.

The aspects (b) and (c) are, in the case of this kind of systems, related. Thus, we will discuss them together. The ease with which a person learns how to use a system is undoubtedly very important. In general, people do not like to use devices that are hard to learn and difficult to use. Essentially, a user of the system has to learn how to express words by pointing at selected icons on the device. In the $Minspeak^{TM}$ system, each word can be expressed either directly by selecting one icon on the keyboard, or by selecting a combination of icons. Thus, the user will have to memorize a fairly large amount of combinations of icons to use the system at a reasonable speed. The ease of remembering the mapping between a certain combination of icons and a word is dependent on the intuitiveness of the combination of icons. In other words, the iconic sentence and the corresponding word should be conceptually similar.

The design environment described in this paper can help the designer of an iconic language select the most intuitive iconic sentence to encode a word. By capturing all the possible meanings of an iconic sentence in a systematic and consistent manner, it enables the designer to choose the combination of icons that will most intuitively express a word. With a domain word in mind, the designer can browse through the meanings of different iconic sentences. The sentence selected will be conceptually most similar to the word. This may facilitate the design of iconic languages for augmentative communication devices which will be easy to use and learn. Moreover, using this design methodology will provide on line information which will facilitate modifications of the visual language. An example is given by the visual language dictionary, which contains the logic derivation of the semantic frames for all the iconic sentences.

It is interesting to note that even if the selection of a particular iconic sentence to encode a domain word is biased by the way a particular designer thinks, the proposed design environment still may help a group of designers of an iconic language be more consistent with their encoding. This is because they would all be using the same CD forms, so the system will make consistent suggestions for each selected iconic sentence regardless of the bias of the individual designer.

In the future, a controlled experiment to evaluate the satisfaction of designers of visual languages using the proposed interactive environment would be highly desirable. It would also be interesting to investigate the possibility of using our interactive environment to customize visual languages for individual users. This will mainly entail the specification of appropriate CD forms to accommodate specific characteristics of individual users, but the customization might also regard the similarity function or the layout of icons and iconic sentences.

6   Appendix

We now show the results of the inference module of our system when run on the iconic sentences apple-glow and money-dice. As explained in previous sections, the algorithm first produces a skeleton of the frame for the iconic sentence, then it enriches it by running the CD form module, which simulate the application of iconic operators.

Iconic sentence apple-glow:

Threshold for the frame apple: 0.7
Threshold for the frame glow: 0.7
Semantic Threshold : 0.7

Frames of the icons composing the sentence:

SLOT:	VALUE	Ω
NAME:	apple	
TYPE:	icon	
RHYME:	cattle	0.20
TIME:	winter	0.80
TIME:	third	0.40
TIME:	fall	0.80
SHAPE:	round	0.80
COLOR:	yellow	0.20
COLOR:	red	0.90
COLOR:	green	0.30
LOCATION:	trees	0.70
QUALITY:	tasty	0.60
USE:	to eat	0.90
EMOTION:	hungry	0.50
RECALL:	sin	0.80
RECALL:	food	0.80
RECALL:	discord	0.80
CULTCONV:	physician	0.60
CULTCONV:	new york	0.70
MASS:	small	0.60
COUNT:	single	0.40
IS_CONCR:	fruit	0.80

SLOT:	VALUE	Ω
NAME:	glow	
TYPE:	icon	
RHYME:	son	0.50
TIME:	noon	0.80
TIME:	morning	0.70
TIME:	evening	0.50
SHAPE:	round	0.85
COLOR:	yellow	0.85
LOCATION:	sky	0.75
QUALITY:	warm	0.85
QUALITY:	powerful	0.85
QUALITY:	bright	0.80
USE:	energy	0.50

EMOTION:	good-mood	0.50
RECALL:	smile	0.60
RECALL:	beach	0.60
CULTCONV:	summer	0.80
MASS:	very big	0.70
COUNT:	one	0.60
IS_CONCR:	special star	0.50

Skeleton Frame of the iconic sentence apple-glow:

SLOT:	**VALUE**	Ω
NAME:	apple-glow	
TYPE:	concept	
TIME:	fall	0.335500
TIME:	winter	0.335500
TIME:	morning	0.205494
TIME:	noon	0.234850
SHAPE:	round	0.824375
COLOR:	yellow	0.419375
COLOR:	red	0.377438
LOCATION:	trees	0.293563
LOCATION:	sky	0.220172
QUALITY:	bright	0.234850
QUALITY:	powerful	0.249528
QUALITY:	warm	0.249528
USE:	to eat	0.377438
RECALL:	discord	0.335500
RECALL:	food	0.335500
RECALL:	sin	0.335500
CULTCONV:	new york	0.293563
CULTCONV:	summer	0.234850
MASS:	very big	0.205494
IS_CONCR:	fruit	0.335500

Derivations with CD forms:

TEXTUAL DESCRIPTION:	Ω	**OPERATOR**
sin IN_THE sky	0.75	"CON"
sin IN_THE_CONTEXT_OF sky	0.75	"CON"
discord IN_THE sky	0.75	"CON"

discord IN_THE_CONTEXT_OF sky	0.75	"CON"
food IN_THE sky	0.75	"CON"
food IN_THE_CONTEXT_OF sky	0.75	"CON"
sin IN_THE morning	0.70	"CON"
sin IN_THE_CONTEXT_OF morning	0.70	"CON"
sin IN_THE noon	0.80	"CON"
sin IN_THE_CONTEXT_OF noon	0.80	"CON"
discord IN_THE morning	0.70	"CON"
discord IN_THE_CONTEXT morning	0.70	"CON"
discord IN_THE noon	0.80	"CON"
discord IN_THE_CONTEXT_OF noon	0.80	"CON"
food IN_THE morning	0.70	"CON"
food IN_THE_CONTEXT_OF morning	0.70	"CON"
food IN_THE noon	0.80	"CON"
food IN_THE_CONTEXT_OF noon	0.80	"CON"
fruit IN_THE morning	0.70	"CON"
fruit IN_THE_CONTEXT_OF morning	0.70	"CON"
fruit IN_THE noon	0.80	"CON"
fruit IN_THE_CONTEXT_OF noon	0.80	"CON"
to eat IN_THE sky	0.75	"CON"
to eat IN_THE_CONTEXT_OF sky	0.75	"CON"
to eat IN_THE morning	0.70	"CON"
to eat IN_THE_CONTEXT_OF morning	0.70	"CON"
to eat IN_THE noon	0.80	"CON"
to eat IN_THE_CONTEXT_OF noon	0.80	"CON"
sin ENHANCED_BY very big	0.70	"ENH"
discord ENHANCED_BY very big	0.70	"ENH"
food ENHANCED_BY very big	0.70	"ENH"
to eat ENHANCED_BY very big	0.70	"ENH"

Iconic sentence money-dice:

Threshold for the frame money: 0.7
Threshold for the frame dice: 0.7
Semantic Threshold : 0.7

Frames of the icons composing the sentence.

SLOT:	VALUE	Ω
NAME:	money	
TYPE:	icon	
ABBREV:	$	0.80
SOUND:	dumb	0.30
TIME:	first	0.50

SHAPE:	rectangular	0.50
COLOR:	green	0.90
LOCATION:	bank	0.70
LOCATION:	pocket	0.80
USE:	investment	0.60
USE:	outliving	0.70
EMOTION:	greedy	0.60
RECALL:	poverty	0.50
RECALL:	wealthy	0.70
LINGCONV:	buck	0.50
MASS:	small	0.20
COUNT:	many	0.60
IS_ABSTR:	debits	0.70
IS_CONCR:	money unit	0.60

SLOT:	**VALUE**	*Ω*
NAME:	dice	
TYPE:	icon	
SOUND:	creak	0.30
TIME:	simultaneus	0.50
SHAPE:	cubic	0.90
COLOR:	black-white	0.70
LOCATION:	casino	0.70
LOCATION:	table	0.50
USE:	gambling	0.70
USE:	game	0.80
EMOTION:	anxious	0.70
RECALL:	fortune	0.70
RECALL:	las vegas	0.60
CULTCONV:	fortune	0.85
MASS:	small	0.40
COUNT:	two	0.70
IS_ABSTR:	ruin	0.50
IS_CONCR:	toys	0.40

Skeleton Frame of the iconic sentence money-dice.

SLOT:	**VALUE**	*Ω*
NAME:	money-dice	
TYPE:	concept	

ABBREV.:	$	0.232000
SHAPE:	cubic	0.182700
COLOR:	green	0.261000
COLOR:	black-white	0.142100
LOCATION:	bank	0.203000
LOCATION:	pocket	0.232000
LOCATION:	casino	0.142100
USE:	outliving	0.203000
USE:	gambling	0.142100
USE:	game	0.162400
EMOTION:	anxious	0.142100
RECALL:	wealthy	0.203000
RECALL:	fortune	0.142100
CULTCONV:	fortune	0.172550
MASS:	small	0.290000
COUNT:	two	0.142100
IS_ABSTR:	debits	0.203000

Derivations with CD forms.

TEXTUAL DESCRIPTION:	Ω	OPERATOR
wealthy IN_THE casino	0.70	"CON"
wealthy IN_THE_CONTEXT_OF casino	0.70	"CON"
outliving IN_THE casino	0.70	"CON"
outliving IN_THE_CONTEXT_OF casino	0.70	"CON"
wealthy ENHANCED_BY fortune	0.70	"ENH"
wealthy ENHANCED_BY two	0.70	"ENH"
wealthy AND fortune	0.70	"COM"

References

1. F. Clarke and I. Ekeland. Nonlinear oscillations and boundary-value problems for Hamiltonian systems. *Arch. Rat. Mech. Anal.* **78** (1982) 315–333.
2. B.R. Baker. Minspeak, a semantic compaction system that makes self-expression easier for communicatively disabled individuals. *Byte*, volume 7, number 9, pages 186-202, September 1982.
3. B.R. Baker and R.F. Barry. A Mathematical Model of Minspeak. 2nd Annual European Minspeak Conference, May 31-June 1, 1990.
4. B.R. Baker and E. Nyberg. Semantic compaction: a basic technology for artificial intelligence in AAC. 5th Annual Minspeak Conference, Nov 14-15, 1990.

5. B.R. Baker, P.J. Schwartz, and R.V. Conti. Minspeak, models and semantic relationships. Fifth Annual Minspeak Conference Proceedings, Seattle, WA, November 1990.

6. C.K. Bliss. Semantography. Sydney: Semantography-Blissymbolics Publications, 1942.

7. N.W. Bray. A cognitive model for Minspeak. 5th Annual Minspeak Conference, November 14-15, 1990.

8. S.K. Chang, G. Costagliola, S. Orefice, G. Polese, and B.R. Baker. A methodology for iconic language design with application to augmentative communication. In *Proc. of 1992 IEEE Workshop on Visual Languages*, September 15-18, Seattle, Washington, USA, pages 110-116.

9. S.K. Chang. Icon semantics - a formal approach to icon system design. *International Journal of Pattern Recognition and Artificial Intelligence*, volume 1, number 1, pages 103-120, 1987.

10. S.K. Chang. Principles of Visual Programming Systems. Prentice-Hall, 1990.

11. S.K. Chang. Principles of Pictorial Information Systems Design. Prentice-Hall, 1991.

12. S.K. Chang. Extending visual languages for multimedia. *IEEE Multimedia Magazine*, Fall 1996, Vol. 3, No. 3, pp. 18-26.

13. S.K. Chang, G. Polese, S. Orefice, and M. Tucci. A methodology and interactive environment for iconic language design *International Journal of Human Computer Interaction*, Vol. 41, 1994, pp. 683-716.

14. S.K. Chang, G. Polese, R. Thomas , and S. Das. A visual language for authorization modeling. In *Proceedings of the IEEE Symposium on Visual Languages*, 1997.

15. G. Costagliola, A. De Lucia, S. Orefice, and G. Tortora. Automatic generation of visual programming environments. *IEEE Computer*, n. 28, pp. 56-66, 1995.

16. R.C. Schank. Conceptual dependency: a theory of natural language understanding. *Cognitive Psycology*, 1972.

17. R.C. Schank. Identification of conceptualization underlying natural language. R. C. Schank and K. M. Colby (eds), W. H. Freeman, San Francisco, CA, 1973.

18. K. Wittenburg. Earley-style parsing for relational grammars. In *Proceedings of the 1992 Workshop on Visual Languages*, pp. 192-199.

Lexicon for Computer Translation of American Sign Language

Robbin Nicole Chapman

Massachusetts Institute of Technology
Artificial Intelligence Laboratory
545 Technology Square, NE43-751
Cambridge, MA 02139
U.S.A.
rnc@ai.mit.edu

Abstract. This work presents a method for translation of American Sign Language (ASL) to English using a feature-based lexicon, designed to exploit ASL's phonology by searching the lexicon for the sign's manual and non-manual information. Manual sign information consists of phonemes *sig* (movement), *tab* (location), *dez* (handshape), and *ori* (hand orientation), which we use as the ASL unit of analysis. Non-manual sign information consists of specific facial and body configurations. A camera acquires the sign and individual frames are analyzed and values assigned to the *sig, tab, dez,* and *ori* sign parameters as well as other sign features, for referencing during lexical search. ASL formational constraints are exploited to target specific image segments for analysis and linguistic constraints serve to further reduce the lexical search space. Primary keys for lexical search are *sig* and *tab*, the most discriminating sign features, followed by the remaining features, as necessary, until a single lexical entry is extracted from the lexicon. If a single lexical candidate cannot be determined, an exception is raised, signaling search failure. This method of using ASL phonological constraints to aid image analysis and lexical search process simplifies the task of sign identification.

1 Introduction

ASL is the visual-gestural language used by Deaf Americans and as such exhibits is own unique phonology, syntax, and semantics, as well as using gesture as primary mode of production. Assistive Technologies applied to language translation can further enhance the quality of life for the Deaf community. This technology will be useful for those who desire alternative communicative means with the non-signing general population. Developing a spatial lexicon to be used for translation between a spoken and visual communication systems presents a challenge due to the differing natures of spoken and visual languages, both in their production and expressiveness. The differing natures of spoken and visual languages arise from the former's production being constrained to a predominately

V. O. Mittal et al. (Eds.): Assistive Technology and AI, LNAI 1458, pp. 33–49, 1998.
© Springer-Verlag Berlin Heidelberg 1998

serial stream of phonemes, whereas sign language phonemes (*sig*, *tab*, *dez*, and *ori*) are generally expressed simultaneously. These sign language peculiarities reveal a lack of isomorphism between ASL and English. There are currently several methods of sign language recognition under development. They range from tethered-glove devices (Starner91) to minimal point-light systems (Bellugi89). The former is being developed at MIT and several other laboratories with hopes of becoming an independent sign acquisition tool for use in practical applications. The latter, point-light display, is being used as an ASL research tool and not for use in assistive technologies.

2   ASL Linguistic Properties

ASL word meaning is conveyed through the sign's linguistic components, which can be described by four parameters, *tab*, *dez*, *sig*, and *ori*. *Tab* refers to sign location with respect to the signer's body. *Dez* indicates sign handshape. *Sig* notates the movement involved in producing the sign. *Ori* notates sign orientation, indicating what part of the hand points toward the ground. Most signs are produced by executing the given specifications of each phoneme category simultaneously. When discussing the physical properties of spoken languages, we refer to such terms as pitch, stress, and modulation. ASL conveys these properties through facial expression, head and body movement and space usage in relation to the signer.

2.1   Phonology.

Phonology typically refers to the patterning of sound, with vowels and consonants comprising these patterns for English. ASL phonology has been defined as the level of systematic-formational structure that deals with sign production, the sub-morphemic units that combine to form each unit, including the restrictions and alternations among these combos (Battison74). ASL exhibits this structure at the sub-lexical and phonological level through patterning of the sign's formational units, with restrictions on how these units are organized. Phonological distinctions are made via hand configuration, place of articulation, and movement. ASL contrasts approximately nineteen handshapes, twenty-four movement paths (i.e., forward, up-and-down, rightward), twelve sign locations, and five base hand orientations (Klima79). *Sig* and *tab* are key lexical items in the identification process, as they have been shown to lead to sign identification more quickly than the other three parameters, due primarily to specific constraints on their valid successors (Emmorey93).

2.2   Formative Constraints.

Phonological constraints on sign formation define legal motion combinations and relationships between the hands. *Symmetry Condition* is evident in two-handed

signs with both hands moving, the constraint requiring handshape and movement specifications to be identical and symmetric. *Dominance Condition* applies
to two-handed sign with two different handshapes. This constraint allows one
hand (usually the dominant) to be moving, with six possible valid handshapes
in the non-moving (or base) position of the sign (Shaepard-Kegl85). Other constraints include: 1) no signs with more than two movements, 2) initial or final *dez*
constrained to six possible handshapes, 3) sign sequences within a sign involving
change in one feature only, 4) majority of body contact areas constrained to
trunk, head, arm, or hand, and 5) horizontally located signs being expressed in
sequence.

Table 1. Examples of ASL Formative Constraints

Parameter	Value	Constraint
SIG	two-handed signs	With interacting hands, movement is frequently alternating along one of the sign planes.
	two-handed signs	Approaching hands and the separation are always on width axis.
DEZ	G-classifier	Most frequently occuring handshape. No variants.
	two-handed signs	If not same *dez* as dominant hand, must be 5, O, A, S, or G handshape only.
	4, F, H, bent-H	Rare occurrences.
TAB	hand	Hand as second *tab* not permitted.
	under-chin	D-classifier handshape only
	center-of-cheek	B-handshape only.
	side-of-cheek	A-thumb handshape only
	upper-arm	H, 4, or L handshapes only.

2.3   Linguistic Constraints.

ASL uses a combination of manual features (MF) and non-manual features
(NMF) to convey meaning. Both have linguistic constraints associated with their
use. Allowable instances of touching, brushing, handshape changes, and location
sequences have constraints which characterize the domain of the ASL syllable
(herein referred to as phoneme). Touching and brushing involve the movement
of one hand in contact with the other hand or with the body. Brushing occurs
at very specific instances, with co-occurrence constraints on handshape changes.
Allowable sequences of locations are constrained within non-compound signs,

with locations differing within a single body dimension only (Sandler89). This is similar to English syllables having certain configurations and constraints on how they can be put together.

Spoken languages distinguish phonologically between nouns and verbs, via inflection classes. ASL distinguishes nouns from verbs on two levels, one with the noun-verb as a purely lexical item, the other with the noun and verb relation to one another (Supalla76). Many nouns and verbs that are lexically and conceptually related in English (i.e., FISH[1] (n) or FISH (v) do not always share that relationship in ASL, each having a different and distinct sign). Relational nouns refer to objects which relate to a verb for the action performed on that object. A classic example is the sign for CHAIR (n) and SIT (v). In context, SIT refers to a chair. In these cases, variance in movement, specifically the noun movement being restricted, disambiguates between the noun and verb form. Noun-verb pair distinctions are also made through re-duplication of the noun base form. An example is the noun-verb pair DECISION/DECIDE, where the noun base form is repeated (DECISION-DECISION) to distinguish it from the verb form (Supalla76). There are many other linguistic and formative constraints in ASL, many of which are being examined for their usefulness in sign prediction and identification from feature values and constraint information.

Pronoun references are made through a variety of methods including indexing, marking, and incorporation. Potentially an infinite number of pronoun references can be made, and they refer unambiguously to a specific referent. Person is derived through its location as referenced in the signing space. If indexed with the conversant, it becomes the second person, if with the signer, it becomes the first person, and any other position, it becomes third person. Marking involves having the signer's body take on referent identity. The signer's body may establish the location of reference by turning to face a particular location and sign a nominal or name in that position. The signer may also assume referent identity for a particular location by occupying that location. Incorporation involves a pronoun configuration (consisting of at most one formal parameter) being incorporated into other signs. Currently, neither marking or incorporation are being handled by the lexicon.

2.4   Grammar and Spatial Syntax

The grammar of a language provides a set of mechanisms that can be used to convey the semantic relations among lexical units required for understanding what is said. Non-manual features begin sentences, change topics, link compounds, state questions, etc., and are conveyed through facial expression, posture, and body position. Non-manual signals are important in ASL and serve many roles, including holding scope over lexical, phrasal, or clausal levels. Non-manual features associated with particular lexical items occur only while the sign is being

---

[1] Herein, uppercase words denote the actual sign, not the written word.

articulated. Features with phrasal domain become part of the phrase grammar. There are also non-manual signals which are grammatical markers (i.e., for relative clauses and adverbials). Clearly, the non-manual signal has lexical scope. Currently, non-manual signals are accommodated in the lexicon structure, but are not included in lexical search and image processing.

3   Constructing the Lexicon

To take advantage of the numerous ASL constraints, the lexicon entry needs to capture a sign's spatial characteristics and contain information about sign meaning, lexical category, and manual and non-manual features.

3.1   Lexicon Notation

Table 2. Lexicon Format

word:	lexical item
c:	item category (noun, verb, pronoun, adverb, etc.)
MF:	manual feature list consists of *sig*, *tab*, *dez*, and *ori* information, number of hands, and hand symmetry, number of sign repetitions, and movement restriction, if any.
NMF:	non-manual feature list contains information about eyes, brow, cheeks, head tilt, mouth, eyes, tongue, and nose configurations.

The feature-based lexicon format developed for ASL uses a similar notational structure as Sandiway Fong's Pappi Principles and Parameters Parser (Fong91). It is essentially a list of lists specifically structured to accommodate ASL features. Some features have default values depending on their category (i.e., the restrictive feature, denotes sign articulation using smaller movements within the sign space. Nouns have restricted movement (-), verbs do not(+)).

3.2   Sign Coordinate System

Due to its spatial properties, sign movement occurs along one of three planes, organized as a three-axis coordinate system. The x-axis, which extends outward from the signer in a forward direction, is labeled the depth-axis. It consists of

Table 3. Manual Sign Features

Notation	Description and Legal Values
tab(dom[],weak[])	Sign location in relation to signer's body. Dom[] indicates dominate hand. Weak[] indicates non-dominate hand. Double instances of dom[] and weak[] are possible for complex signs. Possible values are: top, forehead, eyes, nose, ear, mouth, cheek, neck, chest, stomach, upper-arm, lower-arm, chin, and hand.
sig(dom[],weak[])	Sign movement in relation to signer's body. Possible values are toward (body), forward, downward, upward, side (side-to-side), rotating, circular. No entry means the hands stay in same location.
dez(dom[],weak[])	Sign handshape. Any classifier handshapes.
ori(dom[],weak[])	Sign orientation indicates which part of hand is facing the floor. Possible values are palm (p), pinky (k), back of hand (b), and no orientation. No orientation indicates the hand is held vertical. In this case orientation is in relation the palm direction. Possible values are palm facing body (f), palm facing outward (o), palm facing leftward (l), and palm facing sides of body (s).
hands(num(1,2),sym(±))	Number of hands used in sign and if symmetry.
freq(±)	±Repetition of sign movement.
rest(±)	±Restrictive sign movement.
contact(+/-,(what,how,where))	If contact is made with other hand or body part, what is the hand initiating the contact, how is contact made, and where the contact is made.

characteristic movements such as to-and-fro, toward (the signer), and away (from the signer). The y-axis, labeled width-axis, consists of movements along the horizontal plane, such as rightward, leftward, and side-to-side. The z-axis, labeled vertical-axis, consist of movements along the vertical plane, such as upward, downward, and up-and-down. Within each plane, the movements categories are mutually exclusive.

3.3   Organization of Sign Location

Sign locations are organized in layers, from gross to detailed. Gross features give more general information about a feature (i.e., more round, more straight, spread out, etc.). When analysis indicates an open-handshape is probable, even if further processing is necessary to determine *which* open-handshape, the search routine makes use of this information, using open-handshape as a search key. For many signs, this may be enough information, in conjunction with other

Table 4. Non-manual Sign Features

Notation	Meaning	Legal Values
eye[]	Eye configuration.	normal, wide, and slits. Note, there are many possible eye configurations, but we are starting with this small subset.
cheeks[]	Cheek configuration.	puffed, (sucked) in and normal .
brow[]	Brow configuration.	raised, furrow, and normal.
tongue(±)	Tongue configuration.	regular (-) and stuck out (+).
nose[]	Nose configuration.	regular, twitch, and in.
mouth[]	Mouth configuration.	o-shape (o), closed (c), and curled upper lip (l). Note, there are many possible mouth configurations, but we are starting with this small subset.
head[]	Head tilt configuration.	up, down, left, right.

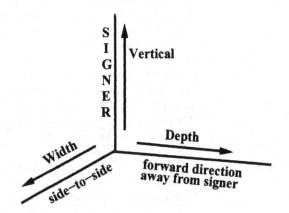

feature information, to identify a single lexical candidate. If not, as more detailed handshape information is made available, further reduction of the candidate subset occurs. For example, if the *tab* is determined to be on level 1 (entire head area), this information, in conjunction with any constraint processing, could be helpful in assigning a feature value. The image segment analysis continues to try and determine a more precise value for the feature. If this precise information is found (i.e., level 1.c (lower-face area)) the search routine uses the this more precise data to further reduce the lexical search space. This process of refinement continues until the lowest level of detail for the region is determined, if necessary.

Lexicon Entry Examples. The general lexicon entry format is:

```
entry(word,c
MF[sig(dom[],weak[]), tab(dom[],weak[]), dez(dom[],weak[]),
ori(dom[],weak[]), hands(number,sym(+/-)),freq(+/-),rest(+/-),
contact((+/-),(what,how),(where))], NMF[eye(), cheeks(), brow(),
tongue(), nose(), mouth(), head()]);
```

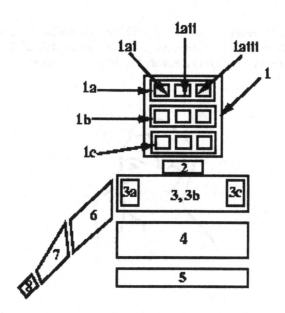

1. head
   (a) upper-face (forehead, temple)
      i. right-temple, forehead
      ii. forehead
      iii. left-temple, forehead
   (b) mid-face (nose, cheeks)
      i. right cheek
      ii. nose
      iii. left cheek
   (c) lower-face   (mouth,   chin,
       cheeks)
      i. right cheek
      ii. mouth, chin
      iii. left cheek

2. neck
3. chest (chest, shoulders)
   (a) right-shoulder
   (b) chest
   (c) left-shoulder
4. stomach
5. waist
6. upper-arm and elbow
7. lower-arm
8. hand

The following *dez* values are used in the next four examples[2].
- b-CL   flat open hand with fingers together; represents a flat sur-
              face or object.
- c-CL   cupped hand; represents curved surface or rimmed object.
- open-o-CL   o-handshape, with middle, ring, and pinky fingers spread
              and extended.
- tips-CL   tips of all fingers and thumb meet at each point.

4   Search

Entries are arranged in the lexicon by category and within category by word. *Tab*
is the first search parameter because it is more readily extracted from the image

---

[2] All sign diagrams and descriptions from [12].

```
entry(eat,v,
MF[sig(dom[toward],weak[]), tab(dom[chin],weak[],dom[mouth],weak[]),
dez(dom[tips-CL],weak[]), ori(dom[p],weak[]), hands(1,sym(-)),freq(+),
rest(-),contact(-)], NMF[eye(), cheeks(), brow(), tongue(), noso(),
mouth(),head()]);
```

Fig. 1. Lexicon entry for the verb *eat*. The closed right hand goes through the natural motion of placing food in the mouth. This movement is repeated.

than the remaining morphemes and for a subset of signs this information, coupled with the sign's *sig*, will lead to a match. The sign start and end *tab* information initiates the search process, although minimally, the start *tab* information is sufficient. The next feature searched is *sig*, whose start and end position data is already present and help to derive the plane the movement occurs in. Next, the *sig* value is determined and applicable linguistic constraints further narrow the candidate subset. This additional information aids in image analysis of *dez* and *ori*, the most difficult information to obtain. Other manual features, such as number of hands, hand symmetry, frequency, and restriction are retrieved and determined, as needed.

Exploiting Constraints. There are ASL linguistic constraints, a few mentioned earlier in this paper, which help predict how or where a feature may appear in the sign stream and to shrink the search space of lexical candidates. One example is the constraint on touching. If an initial *tab* is involved touching in a particular region, the next *tab*, if any, is constrained to the same region plane. For example, if the sign involved touching under the right eye, the only valid next *tab* for the remainder of the sign will be the area under the left eye. Therefore, that region of frame four is examined first. An example of search speedup, not associated with the primary sign features (*sig, tab, dez, ori*), is information about the number of hands involved in the sign. If the weak hand is noted as being beside the body (out of the sign space), this indicates a single-handed sign. If the weak hand is missing from that location, then we are dealing with a

```
entry(book,n,
MF[sig(dom[],weak[]), tab(dom[chest],weak[chest]),
dez(dom[b-CL],weak[b-CL],dom[b-CL],weak[b-CL]),
ori(dom[k],weak[k],dom[b],weak[b]), hands(2,sym(+)), freq(-), rest(+),
contact(-)], NMF[eye(), cheeks(), brow(), tongue(), nose(),
mouth(), head()]);
```

Fig. 2. Lexicon entry for two-handed, symmetrical sign *book*. The open hands are held together, fingers pointing away from body. They open with the little fingers remaining in contact, as in the opening of a book.

two-handed sign and this information will eliminate more than sixty percent of the lexicon. If symmetry status can be determined, this will further shrink the size of the subset of lexicon candidates. There are many such shortcuts possible by exploiting ASL linguistic constraints.

4.1 Processing Hazards

Image feature extraction and subsequent value assignment can lead to hazards which jeopardize successful lexical search and parsing. These hazards fall under three categories:

Feature Incompleteness. This hazard involves a feature not being recognized by the system and therefore not given a value. The signs for KNIFE and EN-THUSIASTIC present an example of this hazard. Both are two-handed signs and one may be inadvertently identified as the other if the search algorithm overlooks the non-dominant (weak) hand movement during image processing. While both signs have forward movement of the dominate hand, ENTHUSIASTIC also has a simultaneous backward movement of the non-dominant hand. If this feature is missed, and no value is assigned to the non-dominant *sig* feature, the incorrect lexicon entry KNIFE would be selected instead of ENTHUSIASTIC. "Knife" is a noun and "enthusiastic" an adverb. With the noun occupying the adverbial position in the output string, the parse attempt will fail.

Feature Mismatch. This processing hazard involves a feature being assigned an incorrect value. Feature mislabeling could occur as a result of the similarity of properties for a particular morpheme (i.e., the *sig* values [arc-forward] and [forward] are very similar). For example, if the sign GIVE, which has the label *sig*=[arc-forward], was instead assigned the label *sig*=[forward], search would find

```
entry(buy,n,
MF[sig(dom[],weak[],dom[small-arc-forward],weak[]),
tab(dom[chest],weak[chest]),
dez(dom[c-CL],weak[b-CL]), ori(dom[b],weak[b]), hands(2,sym(-)), freq(-),
rest(+), contact(+,dom[hand,atop],weak[palm])], NMF[eye(), cheeks(),
brow(), tongue(), nose(), mouth(), head()]);
```

Fig. 3. Lexicon entry for simple two-handed, non-symmetrical sign *buy*. The upturned right hand, with thumb closed over palm of hand, is brought down onto the upturned left palm. The right hand moves forward and up in a small arc, opening as it does.

a match with the entry LIKE, which has *sig*=[forward] and shares the same *tab* as GIVE. Further processing of the remaining morphemes (*dez* and *ori*), which are not the same in both signs, would result in an exception marker being raised, indicating that no lexical entry was found. In this case, an existing lexicon entry has failed to be identified during the lexicon search because of an incorrectly labeled feature. A possible solution to this hazard would be to keep a table of similar feature values, for substitution when a lexical search failure has occurred, which may allow recovery of the correct lexical entry.

Too Many Matches. Another processing hazard involves the search process returning more than one entry. Usually, when several entries share the same combination of *sig, tab, dez* and *ori*, the remaining manual (hands, freq, rest, and contact) non-manual features are referenced to "break the tie." Of course, if they are not available, possibly due to a feature incompleteness hazard, the result could be too many entries selected with no way to disambiguate them.

Wildcards. Another possible cause of this hazard is the use of wildcards. Wildcards are a provision for assigning values to image segments whose parameter value cannot be determined, while additional image processing continues. In fact, wildcards are a processing property, not a lexical item and necessary for increasing the chance of a lexicon entry being identified, even with incomplete information. In many cases, a single lexical item may be identified with partial information if some subset of its features is unique. However, if this is not the case, the result will be too many entries being selected with no way to disambiguate them. The signs TAPE, CHAIR, and TRAIN demonstrate how wildcards may cause a hazard. These signs are minimal pairs, sharing the same *dez, tab*

```
entry(cat,n,
MF[sig(dom[right],weak[left], dom[back-and-forth],weak[]),
tab(dom[upper-lip],weak[upper-lip], dom[chest],weak[chest]),
dez(dom[open-o-CL],weak[open-o-CL],dom[c-CL],weak[c-CL]),
ori(dom[s],weak[s],dom[p],weak[p]), hands(2,sym(-)), freq(-), rest(+),
contact(+,dom[],weak[],dom[fingers,atop],weak[hand])],NMF[eye(),cheeks(),
brow(), tongue(), nose(), mouth(), head()]);
```

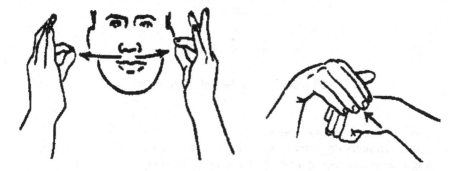

Fig. 4. Lexicon entry for complex two-handed, non-symmetrical sign *cat*. The thumb and index fingers of both hands (open-o-classifier) stroke an imaginary pair of whiskers. The right hand touches the back of the left hand.

and *ori* values, with the *sig* parameter distinguishing them. If *sig* is assigned the wildcard value, there may be no way to select a single lexical entry. This would cause an exception marker to be raised, indicating lexical search failure. A possible solution to this hazard is to use knowledge of ASL linguistic constraints to "guess" a value for the feature. For example, once a wildcard is detected, based on the values of previously identified features, ASL sign formation constraints may be able to assign the most probable value to the feature containing the wildcard value, further updating this probable value as other information is extracted from the image.

5 Using The Lexicon

The lexicon currently consists of two hundred signs representing two hundred English words or phrases. The lexical categories are noun, verb, adjective, adverb, preposition, pronoun, and wh-noun.

5.1 Simplifying Assumptions

There are several simplifying assumptions made for lexicon use, specifically

- different lexical entries sharing the same sign have been omitted from this version of the ASL lexicon

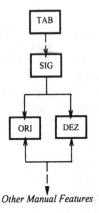

Other Manual Features

Fig. 5. Lexical Search Hierarchy

- fingerspelling is not processed
- noun-verb pairs determined by reduplication are ignored
- single signs expressing a series of words are ignored
- no micro-movement (wiggling, fingerspelling, etc.)
- no body aspect changes (signer is always facing forward)
- hand crossing, which occurs a some signs, is ignored.

5.2 Manual Feature Extraction

With the average sign length of approximately one second, twelve frames are used to capture a single sign. Pairs of frames, indexed at the first, fourth, eighth, and twelfth frame positions, depending on which sign feature is being determined, are examined during the identification process. This particular frame index configuration makes sense for several reasons. One, timing of the first and fourth frames corresponds to the approximate start and end points of the sign. This allows analysis of the *tab* parameter at the start and end of the sign (tab_i and tab_j). This information also conveys the start and end points of the *sig* and, of course, the start and end of *dez* and *ori*. The fourth and eighth frames provide additional point information about the location of the moving hand, which can help determine the path of the movement (Klima79). In this way complete *sig* information is extracted from the frame sequence. Sign boundaries are currently delimited by *dez* and *tab* change, but there are numerous NMF cues for sign boundaries that will be incorporated in later editions. The central signing space for one-handed signs is the front of the neck, two-handed signs tend to lower toward the chest area. If no value, including gross values, can be determined, the wild card value (*) is assigned to the applicable feature and passed to the search routine. When no signals have been detected, the parse routine is activated and the output string is parsed using the Earley parsing algorithm (Earley68), with the resulting parse tree format: ((ROOT (S (*applicable lexicon entries preceded by their syntactic labels...*)))).

5.3 Example

Consider the ASL utterance, "ix-1st-person[3] buy book." Non-manual linguistic information is ignored in this example, therefore lexical tense markers cannot be determined. We will assume present tense. Three signs are recorded and the images organized into three 12-frame units.

1. The first 12-frame sequence consists of signer pointing to herself with the index finger at chest level, indicating the ix-first -person pronoun.
2. The next 12-frame sequence is examined. *Tab* and *sig* are established, with the search starting with whichever feature is first available. If no single lexical candidate emerges from the search, *dez*, *ori*, and other manual features are used as they become available until a single entry, in this case 'buy,' is identified and appended to the output string.
3. Finally, the last sign, "book," undergoes a similar process, with *tab, sig* and *number* being the distinguishing features for this sign although, in our two hundred word lexicon, having the *dez* and *ori* values would negate the need for knowing the number of hands used to produce the sign.

4. When the signer's hands rest by her sides, end of utterance is detected and the string is forwarded to the parsing algorithm which produces the following output:

((ROOT (S3 (PRO xi-1st-person) (VP (V BUY) (NP (N BOOK)))))).[4]

6 ASL Translation and Assistive Technology

While projects are underway elsewhere, there is still a need for more study of ASL translation and its application to assistive technologies. When faced with an activity in a given context (business meeting, emergency situations, educational environment, etc.), the Deaf individual may utilize assistive technologies (AT) to enhance their ability to carry out a functional activity (Cook95). This definition emphasizes functional capabilities of individuals with disabilities, with functional outcomes being the measure of success of AT technology. Assistive technologies for the Deaf bring with it issues of portability, usability, and versatility. An ASL-to-English translator would be invaluable in the workplace, schools, and for use in emergency situations. These more essential uses are driving the development of this technology.

---

[3] ix-1st-person is lexical notation for the first person pronoun. This is physically indicated by the signer pointing to herself at chest level.

[4] This phrase happens to be in SVO order, however, ASL also uses SV and SOV word orderings, as well.

6.1   User Interface

While the ability to translate a sign language to a spoken language wouldaid to the Deaf community and general population in communicating together more freely. Seamless user interaction with this technology is just as vital. This inter-action requires a user/system interface, which represents a boundary between the user and the assistive technology. Such a GUI would guide the user through system initialization processes, such as camera calibration and conversant names for dialog labeling. The system must also be capable of handling input from the hearing users of the system (i.e., requests for signer to repeat themselves, etc.) as well. Of course, the GUI must ultimately be easy to navigate.

6.2   Determining Community Needs

Part of the motivation for trying to build a translation system using a camera for sign acquisition is to meet the needs of the primary user, the Deaf individual. The assimilation of an assistive technology will be highly determined by its comfort level, ease of use, appearance, and how it makes the user look and feel about themselves. To date, investigation into these issues has been mainly informal conversations with Deaf students at MIT and Boston University regarding what they would consider unacceptable conditions for using the technology, what fea-tures would they like to see, etc. The general consensus from this small group of eleven is that they don't want to anything that will call undue attention, causing them to feel overly self-conscious. They also want control over the output mode, speech or text. Surprisingly, almost all wanted to chose the gender of the audio output, if possible. Clearly, in enhancing the Deaf individual's communication with the general population, we must be concerned with the social context in which this performance takes place. A formal study is required to determine which sections of the community would make use of this technology and any configuration and operational requirements they deem necessary as part of an ASL translation system.

7   Conclusion

7.1   Discussion

Our lexicon structure is designed to exploit ASL phonology for sign identifica-tion. The real value of this feature-based lexicon is it allows search for entries by the sign structure, which is a more manageable way of encoding ASL's three-dimensional qualities. While the idea of encoding ASL features is not new, pre-vious efforts, mostly linguistically driven, have attempted to encode a great deal of sign detail. I am attempting to encode only what is necessary for sign identi-fication using the technology available. By not requiring all the linguistic detail, the goal is to make the sign recognition problem more tractable. An example of this approach is my encoding of the *ori*, or hand orientation, parameter. Rather than require *ori* sub-features for finger placement and direction, in addition to palm orientation, only palm information is included. The palm information is considered to be more reasonably extracted from the image, being a grosser feature than finger direction, and coupled with other parameter constraints (i.e., particular *dez* or *tab* configurations with particular orientations) help discriminate between signs having distinctive finger orientation. Understanding of ASL structure and constraints facilitate sign identification from this abbreviated view of the sign. Additionally, rather than analyzing the entire image to ascertain sign meaning, a sign's features are pieced together, like a puzzle, to identify the sign. This approach allows for system robustness because sign identification is derived from its significant features and, as such, variances in sign formation will be more easily handled. Second, the morpheme identification versus full image analysis effectively removes the bulk of responsibility for sign recognition from the visual system. Analysis of an image segment involves less processing than identification an entire image. Third, in actual use this method of searching the lexicon may speed up the retrieval process. Once a sign feature search selects a group of candidates only those candidates are checked on in future search iterations. Exploiting ASL linguistic and sign structure constraints also aid in sign identification. Many of these factors allow the sign to be identified from only a portion of the image.

7.2   Looking Ahead

The sign identification process is a challenging slice of the ASL-to-English translation pie and will determine the reliability of such a translation system. This type of assistive technology would be invaluable in the workplace, schools, and for use in emergency situations. Of course, to use this type of system in everyday environments introduces its own unique problems, such as signal noise and variance in lighting and background, and semantic segmentation are all important and must be addressed before it can be put to practical use. Feature enhancement of the lexicon introduced in this text will help the sign identification process by allowing for alternate subsets of features to unambiguously identify specific signs. Also, more aggressive determination of sign boundaries and the incorporation of additional formative and linguistic constraints to aid identification efforts are necessary to enhance system robustness and versatility.

Acknowledgments

This research is supported by a Ford Foundation Fellowship administered by the National Research Council.

References

1. R.H. Battison. Phonological deletion in ASL. *Sign Language Studies* 5(1974):1-19, 1974.
2. A. Cook. *Assistive Technologies: Principles and Practice*. Mosley-Year Book, Inc., 1995.
3. B. Dorr. *Machine Translation. A View from the Lexicon*. The MIT Press, 1993
4. J. Earley An efficient context-free parsing algorithm. Ph.D. dissertation, Department of Computer Science, Carnegie-Mellon University, 1968.
5. K. Emmorey. Processing a dynamic visual-spatial language: psycholinguistic studies of american sign language. *Journal of Psycholinguistic Research* 22(2):153-183, 1993.
6. S. Fong. Computational properties of principle-based grammatical theories. Ph.D. dissertation, Dept. of Computer Science, Massachusetts Institute of Technology, 1991.
7. E. Klima and U. Bellugi. *The Signs of Language*. Harvard University Press, 1979.
8. S. Liddell. *American Sign Language Syntax*. Mouton Publishers, 1980.
9. D. Lillo-Martin. *Universal Grammar and American Sign Language*. Kluwer Academic Publishers, 1991.
10. J. Shepard-Kegl. Locative relations in american sign language word formation, syntax, and discourse. Ph.D. dissertation, Dept. of Computer Science, Massachusetts Institute of Technology, 1985.
11. T. Starner, T. Visual recognition of american sign language using hidden markov models. S.M. thesis, Massachusetts Institute of Technology, 1991.
12. M. Sternberg. *American Sign Language Dictionary*. HarperCollins Publisher, Inc., 1987.
13. M. Sternberg. *American Sign Language Dictionary on CD-ROM*. HarperCollins Publisher, Inc., 1994.
14. T. Supalla and E. Newport. *Understanding Language through Sign Language Research: The Derivation of Nouns and Verbs in American Sign Language*. Academic Press, Inc., 1976.
15. W. Sandler. *Phonological Representation of the Sign. Linearity and Nonlinearity in American Sign Language*. Foris Publications Holland, 1989.
16. W. Stokoe, D. Caterline, and C. Croneberg. *Dictionary of American Sign Language*. Gallaudet College, Washington, D.C. Linstock Press, Silver Springs, MD, 1978.

On Building Intelligence into EagleEyes

James Gips

Computer Science Department
Boston College
Chestnut Hill, MA 02167
U.S.A.
gips@bc.edu

Abstract. EAGLEEYES is a system that allows the user to control the computer through electrodes placed on the head. For people without disabilities it takes 15 to 30 minutes to learn to control the cursor sufficiently to spell out a message with an onscreen keyboard. We currently are working with two dozen children with profound disabilities to teach them to use EAGLEEYES to control computer software for entertainment, communication, and education. We have had some dramatic successes.

1 The EagleEyes System

EAGLEEYES [4,5,6] is a technology that allows a person to control the computer through five electrodes placed on the head. An electrode is placed an inch above the right eye and another an inch below the right eye. Electrodes are placed at the temples, an inch to the left and right of the eyes. A fifth electrode is placed on the user's forehead or ear to serve as a reference ground. The leads from these electrodes are connected to two differential electrophysiological amplifiers, which amplify the signals by a factor of 10,000. The amplifier outputs are connected to a signal acquisition system for a Macintosh or Windows computer. Custom software interprets the two signals and translates them into cursor (mouse pointer) coordinates on the computer screen.

The difference between the voltages of the electrodes above and below the eye is used to control the vertical position of the cursor. The difference between the voltages of the electrodes to the left and right of the eyes is used to control the horizontal position of the cursor.

The dominant signal sensed through the electrodes is the EOG, or electrooculographic potential, which also is known as the ENG or electronystagmographic potential. The EOG / ENG has been investigated for over 70 years [21]. The EOG/ENG results from the variation in the standing potential between the retina and the cornea [16]. The signal corresponds to the angle of the eye relative to the head. Currently the major use of the EOG/ENG is in diagnosing vestibular and balance problems [3]. In the 1960's and 1970's people experimented with the EOG as a means of determining where people are looking [25]. Currently most approaches to sensing point of gaze use an infrared-sensitive camera or imaging system to visually track features of the eye and then a computer or

V. O. Mittal et al. (Eds.): Assistive Technology and AI, LNAI 1458, pp. 50–58, 1998.
© Springer-Verlag Berlin Heidelberg 1998

some electronics to do the reverse geometry to determine where the user is look-ing [19,8]. Baluja and Pomerleau [1] have reported using a neural network to process ambient light video camera images of the eye to determine where the user is looking.

Using electrodes has its problems for tracking gaze. The EOG/ENG signal is a function of the angle of the eye in the head, so the signal can be affected by moving the eyes relative to the head or by moving the head relative to the eyes or by a combination of both. The signal picked up through the electrodes also can be affected by moving your eyelids [2], scrunching your eyes, moving your tongue, and by other facial movements both conscious and unconscious. (There also can be drifts in the signal caused by, for example, reactions between the skin and the electrodes, and interferences in the signal from various equipment and external anomalies. The signals involved are quite small, typically on the order of 100 microvolts.)

In the EAGLEEYES system we are not so much interested in the traditional tracking of eye movements as we are interested in enabling people to control the computer. For us the many ways the user can affect the signal is an advantage rather than a disadvantage. During initial training and skill development people experiment and arrive at their own optimal method of controlling the cursor. It's a semi-conscious skill like riding a bicycle or skiing. Some people move their head a lot. Some move their eyes. Some use their tongues. Many of the people we are working with have such profound disabilities. Whatever works is fine with us. Think of the brain as a neural net! People arrive at some optimal way of controlling the signal but it is not always quite clear how or what is going on.

Control of the computer through EOG also is being investigated in the Eye-Con/Biomuse system [14] and by groups at Shinshu University in Japan [7] and the University of Vienna [22].

The EAGLEEYES system software allows us to run EAGLEEYES with most existing commercial software. Our system software runs in the background. Every 1/60th of a second it springs to life, senses the two values on the A/D converter, translates the values into screen coordinates, and saves them as the official mouse coordinates in the system. An option in the software allows a mouse click to be generated whenever the cursor remains within a settable small radius on the screen for a certain period of time. That is, the user can generate a click by holding the cursor at a spot on the screen for a certain fraction of a second. The software can run in the background with any well-behaved application. Thus, Macintosh or Windows software, whether commercial or custom-developed, can be run by eye control instead of mouse control.

2   Current Systems

We currently have seven EAGLEEYES systems, three in our laboratories, one in the Campus School, one in a satellite facility at the Reeds Collaborative in Middleboro, Mass., and personal systems in the homes of a 13 year old boy and a 15 year old boy, each of whom have spastic quadriplegic cerebral palsy. The

Campus School is a day-time educational facility for 42 students, aged three to twenty-two, who are non-verbal and have multiple impairments. It is part of the School of Education at Boston College and is located on the main university campus. The EAGLEEYES facility at the Campus School is reserved for full-time Campus School students in the morning and for visitors and other students with disabilities from the greater Boston area in the afternoon. Because of increasing demand, we recently opened up the facility in Middleboro, about an hour south of Boston, to provide access to EAGLEEYES to children in the area. EAGLEEYES facilities at other locations are under discussion.

The personal systems were installed in the homes of two young men who have no voluntary control of muscles below the neck, have no reliable "switch sites," and cannot speak. Both have learned to use EAGLEEYES well enough to spell out messages. Both use EAGLEEYES every day for cognitive academic activities in their school programs; they are able to use EAGLEEYES to do their homework [13,15,20].

3   Using EagleEyes

The EAGLEEYES system mainly tracks the EOG, which is proportional to the angle of the eye in the head. Learning to use the EAGLEEYES system is an acquired skill.

A person without disabilities usually requires 15 to 30 minutes to learn to use the system and to become proficient enough to spell out her name using a keyboard displayed on a screen. For a new user we usually explain that the system is measuring mainly the angle of the eye in the head and that the cursor can be moved either by holding the head constant and moving her eyes or by fixing her eyes on a spot in the screen and by moving her head or by some combination of the two. New users practice moving a cursor on a blank screen and then play a simple video game we have developed for training.

In one study we taught twelve undergraduates (mean age: 20) and ten faculty and staff (mean age: 58) to use EAGLEEYES. Subjects had no previous experience with EAGLEEYES. Each session lasted one hour. By the end of the session all but one of the subjects were proficient enough to shoot down 9 or 10 out of 10 aliens in the video game we use for training. Eleven out of twelve undergraduates and eight out of ten faculty and staff became proficient enough to spell out the message "HELLO EAGLE EYES" through the electrodes with an average of under one error per subject.

For people with severe disabilities it can take anywhere from 15 minutes to many months to acquire the control skill to run the system. First of all, the children need to understand that they are controlling the cursor on the screen by moving their eyes. Children who are completely paralyzed from birth are not used to physically controlling anything, much less the cursor on the screen with their eyes. Once the children understand the cause and effect of moving the cursor with their eyes, we help them develop their control skills by having them run various commercial and custom-made software. For example, one program allows them to "paint" with their eyes. Wherever the child moves the cursor colored lines are drawn. At the end of the session we print out the eye paintings on a color printer and give them to their parents to hang up on the refrigerator or to put in a frame on the wall. The children use video games for practice and also multimedia programs we have developed that allow the user to select one of four digitized video clips to be played by looking at one of four opening frames of the clips presented in quadrants on the screen.

At the invitation of the California Pacific Medical Center in San Francisco we recently tried EAGLEEYES with ten people with ALS (Lou Gehrig's disease). In a 60 to 90 minute session six learned to use the system well enough to hit at least 9 out of 10 aliens in our training video game. Five learned to use EAGLEEYES well enough to spell out messages. One gentleman with advanced ALS spelled out the message "THERE IS NO WAY TO THE END OF THE JOURNEY BUT TO TRAVEL THE ROAD THAT LEADS TO IT".

4   The Human-Computer Interface

A major challenge has been the design of the human-computer interface. That is, given the capabilities of people to move their eyes and head to control the signal and given the physical characteristics of the EAGLEEYES system, the amplifiers and so forth, how should the software be designed so that it is easy for people to use? Jacob [10,11] points out important potential benefits and problems of using eye movements to control computers. For example, he discusses the "Midas Touch problem":

> At first, it is empowering to be able simply to look at what you want and have it happen, rather than having to look at it (as you would anyway) and then point and click it with the mouse or otherwise issue a command. Before long, though, it becomes like the Midas Touch. Everywhere you look, another command is activated; you cannot look anywhere without issuing a command. The challenge in building a useful eye tracker interface is to avoid the Midas Touch problem.

Jacob ([10] page 156)

Generally the software we use with EAGLEEYES is controllable by large buttons or clickable areas. The basic issue is accuracy and control. With EAGLEEYES, the user can move the cursor with fair accuracy and can issue a single mouse click by briefly holding the cursor at a spot.

We have experimented with using voluntary blinks instead of dwell time to cause a mouse click. We have written software to detect voluntary blinks versus involuntary blinks. We have found voluntary blinks to be slower and less accurate than dwell time in making selections. (When a person blinks there is a large spike in the vertical EOG. It takes some time for the vertical signal to recover.) With a third pair of electrodes and another amplifier and signal channel we have devised software to detect winks of the left and right eye. Undergraduates can learn to blink quickly twice in succession for a double click or that winking the left eye causes the left mouse button to be depressed until the next wink. This is not very natural. The more fundamental problem with this general approach for us is that we have found that the children and young adults we are working with cannot voluntarily blink or wink their eyes!

We have adapted EAGLEEYES to use with the now standard WIMP (Window-/Icon/Mouse) interface but it does not seem quite right for our users. We are groping towards a better interface to use with EAGLEEYES. Nielsen [17] and Jacob et al. [12] provide interesting discussions of trends for next generation user interfaces, including some possible future roles of eye tracking devices.

We have experimented with two forms of continuous movement through a virtual space with EAGLEEYES.

One form is the movement through a virtual visual space. For example, EAGLEEYES works with classic commercial video games where you (your character) move around through a simulated world. A favorite with some of our older male users is to run commercial flight simulator game programs strictly through the electrodes of EAGLEEYES. Of course you use your eyes to gather information and your eye movements can be involuntary. It's easy for your plane to spin out of control when using a simulator with your eyes, perhaps even more so than with a mouse or joystick.

A second interesting form of continuous movement through a virtual space is to use EAGLEEYES with a real-time musical composition program. Here the space is auditory. The music that emerges from the computer depends on the cursor location. Normally the user composes by moving a mouse or joystick; the music changes in real-time. With EAGLEEYES, the user composes by moving his eyes. Since EAGLEEYES works equally well with eyes closed, you can sit back in your easy chair with your headphones on, close your eyes and relax, and compose music by moving your eyes and head.

A basic trend in human/computer interfaces has been a continuing shortening of the feedback time between human and computer. In the bad old days of punched cards and batch jobs the feedback time was measured in hours or days. Now the feedback time is measured in seconds. Still, there should be a way to shorten it even more by eliminating the voluntary motions of moving the hand to move the mouse and click the button. Shortening the feedback time seems to lead to increased realism and better interfaces.

5   Controlling Mobile Devices

A student project allows a commercial remote-controlled toy car to be controlled through EAGLEEYES. Basically the idea is to look left and the car moves left, look up and the car moves forward, etc. This is a big hit with children who have never been able to control a physical device.

In recent work [24] EAGLEEYES was used to control the Wheelesley robotic wheelchair [23]. A robotic wheelchair has sensors and an internal computer so it can automatically avoid running into obstacles and walls. (See the several papers in this volume.) With EAGLEEYES the driver looks up to move the wheelchair forward. The driver looks left to move the wheelchair to the left. And so on. This can be considered a real-life extension of the use of EAGLEEYES to move through virtual visual spaces. Since the driver is using the eyes to take in information as well as to steer the wheelchair, this takes a certain concentration. The sensors and the internal computer can do most of the work of controlling the chair and keeping it out of trouble. The driver is needed mostly at decision points, for example at hallway intersections or when turns are needed in open spaces. We demonstrated this highly experimental system at the American Association for Artificial Intelligence '97 national conference in Providence, RI. Work is continuing on improving the interface and the control methods. The goal is to create a wheelchair system that people with profound disabilities can use, people who have no chance at becoming competent drivers of conventional powered wheelchairs.

6   Custom Software

We have developed several types of applications software to use with the system – communications software, educational software, entertainment software [5].

In communications software, we have developed a classic "spell 'n speak" keyboard program. We have found the typing speed using EAGLEEYES with a full on-screen keyboard to be about one character every 2.5 seconds for short (three or four word) messages. The accuracy required for a 30 character keyboard is too great for the children with disabilities with whom we are working. We have worked on perhaps a dozen iterations of a two-level system, where the user first selects a group of letters and then selects a letter from the group. We also have worked with various augmented communication   systems, like Boardmaker and Speaking Dynamically, where the user looks at icons and the computer speaks words or phrases that correspond to the icons. Most children with profound disabilities are taught to look up for "Yes" and down for "No". One of the first communications programs we usually try with a child is a program that asks a question (typed in ahead of time by a parent or teacher) and then says "Yes" for the child if the child moves the cursor up and "No" if the child moves the cursor down. Once this skill is mastered we can move on to a spelling program that allows the user to select a letter by looking up for "Yes" and down for "No" in response to questions like "Is the letter in the group ABCD?"

We have developed several types of educational software. One often-used program administers multiple choice tests via eye control. The question is placed in the center and four choices are placed off in the corners (or in a + pattern). This program is being used every day by the two teenagers with EAGLEEYES systems in their homes.

All of this software works with a traditional mouse as well as with EAGLEEYES. It simply is designed to be controlled with large buttons and single clicks and to be as easy to use and transparent as possible. The content is designed to be useful or amusing for the children and young adults with whom we work.

7   Intelligence

Currently the intelligence in the EAGLEEYES system resides in the user. We provide tools and feedback, but it is up to the user to learn how to control the electrical potentials that are sensed by the electrodes on his or her face. The guiding principle in the design of the EAGLEEYES hardware and software has been KISS – Keep It Simple, Stupid.

The EAGLEEYES system and processing is as simple and transparent as possible. The system is self-centering. The only initial calibration we do is to adjust the gain on each channel and that is only on systems that are used by multiple people. On systems in children's homes, there are no initial calibrations or adjustments necessary as the gain settings usually are appropriate from one session of a single user to the next. During the processing either we use the raw signals to control the mouse pointer or we do some simple exponential smoothing on the signals, at the choice of the user.

A question for us is how we might build intelligence into the EAGLEEYES system itself, for example into the "mouse" drivers. EAGLEEYES is an interface between the user and an application program on the screen. We might have EAGLEEYES more aware of the application program or more aware of the user or both.

EAGLEEYES is designed to work with any application program. One approach to making it aware of the application program might be for it to examine the screen, decide what might be the points of interest (for example buttons or, more generally, locations of color discontinuities), and then try to determine if the user is trying to reach one of those points. The program might attempt to assist the user in moving the cursor to those points and perhaps in issuing a mouse click there.

EAGLEEYES receives signals from electrodes placed around the eyes: these correspond primarily to the angle of the eye in the head and are affected by eye movements.

Human eye movements have been studied for many years and much is known about them [25,18]. When scanning a visual scene or a computer screen or reading text the eyes often engage in "jump and rest" behavior. A saccadic eye movement is a rapid ballistic movement by which we voluntarily change fixation from one point to another. A saccadic eye movement can take from 30 to 120 milliseconds. Between saccadic eye movements the eyes often are relatively still as they fixate on an area of the screen for approximately 250 milliseconds (100 to 500 milliseconds) and take in the information in that area. We could attempt to have the EAGLEEYES driver make use of the characteristics of human eye movements in moving the cursor. For example, the driver might use the ballistic characteristics of saccadic movements in better predicting where the user wants to move the cursor on the screen.

Are these feasible? Would these efforts do more harm than good? The concern is that the control of the cursor might become more complex and unpredictable so the user might have more difficulty learning to control the cursor through EAGLEEYES. Eye movement is not the only source for the signal we are sensing. By our efforts we might make EAGLEEYES more difficult to use rather than easier to use.

We have been working to make the user better able to control the signal. We have been working to make and find applications software that is especially EAGLEEYES-friendly. Whether our users would benefit from an attempt to build intelligence into the EAGLEEYES software itself is an important open question for us.

References

1. S. Baluja and D. Pomerleau. Non-Intrusive Gaze Tracking Using Artificial Neural Networks. In *Advances in Neural Information Processing Systems*, J.D. Cowan. G. Tesauro, J. Alspector (eds.), Morgan Kaufmann Publishers, San Francisco, CA. 1994.

2. W. Barry and G. Melvill Jones. Influence of Eyelid Movement Upon Electro-Oculographic Recording of Vertical Eye Movements. Aerospace Medicine. 36(9) 855-858. 1965.

3. J.R. Carl. Principles and Techniques of Electro-oculography. Handbook of Balance Function Testing, G. P. Jacobson, C.W. Newman, and J. M. Kartush (eds.), Mosby Year Book, 69 - 82. 1993.

4. J. Gips, P. DiMattia, F.X. Curran and C.P. Olivieri. Using EAGLEEYES – an Electrodes Based Device for Controlling the Computer with Your Eyes – to Help People with Special Needs. The Fifth International Conference on Computers Helping People with Special Needs (ICCHP '96), Linz, Austria. In J. Klaus; E. Auff; W. Kremser; W. Zagler; (eds.) Interdisciplinary Aspects on Computers Helping People with Special Needs.. Vienna: R. Oldenbourg. 77-83. 1996.

5. J. Gips and C.P. Olivieri. EAGLEEYES: An Eye Control System for Persons with Disabilities. The Eleventh International Conference on Technology and Persons with Disabilities. Los Angeles. 1996. (See www.cs.bc.edu/g̃ips/EagleEyes)

6. J. Gips, C.P. Olivieri and J.J. Tecce. Direct Control of the Computer through Electrodes Placed Around the Eyes. Fifth International Conference on Human Computer Interaction, Orlando, FL. In M.J. Smith and G. Salvendy (eds.) Human-Computer Interaction: Applications and Case Studies. Elsevier. 630-635. 1993.

7. M. Hashimoto, Y. Yonezawa and K. Itoh. New Mouse-Function using Teeth-Chattering and Potential around Eyes for the Physically Challenged. The Fifth International Conference on Computers Helping People with Special Needs (ICCHP '96), Linz, Austria. In J. Klaus; E. Auff; W. Kremser; W. Zagler; (eds.) Interdisciplinary Aspects on Computers Helping People with Special Needs.. Vienna: R. Oldenbourg. 93–98. 1996.

8. T.E. Hutchinson. Computers that Sense Eye Position on the Display. Computer. July 1993. 65, 67. 1993

9. T.E. Hutchinson et al. Human-Computer Interaction Using Eye-Gaze Input. IEEE Transactions on Systems, Man, and Cybernetics. 19(6) 1527-1534. 1989

10. R.J.K. Jacob. The Use of Eye Movements in Human-Computer Interaction Techniques: What You Look at Is What You Get. ACM Transactions on Information Systems. 9(3) 152-169. 1991.

11. R.J.K. Jacob. What You Look at Is What You Get. Computer. July. 65-66. 1993.

12. R.J.K. Jacob, J.L. Leggett, B.A. Myers, and R. Pausch. An Agenda for Human-Computer Interaction Research: Interaction Styles and Input/Output Devices. Behaviour and Information Technology. 12(2) 69-79. 1993.

13. K. Keyes. Witness to a Miracle. Marshfield Reporter. October 19. Pages A1, A3. 1995.

14. Lusted, H.S., Knapp, R.B., and Lloyd, A. Biosignal Processing in Virtual Reality. Third Annual Virtual Reality Conference, San Jose, CA. 1992.

15. Morgan, B. Knowing Michael. Boston College Magazine. September 1996. .30 - 38. 1996.

16. O.H. Mowrer, R.C. Ruch, and N.E. Miller. The Corneo-Retinal Potential Difference as the Basis of the Galvanometric Method of Recording Eye Movements. American Journal of Physiology. 114,423. 1936.

17. J. Nielsen. Noncommand User Interfaces. Communications of the ACM. 36(4) 83-99. 1993.

18. K. Rayner. Eye Movements and Cognitive Processes. In J. M. Findlay at al. (eds.), Eye Movement Research, Elsevier Science, pages 3-22. 1995.

19. R. Razdan and A. Kielar. Eye Tracking for Man/Machine Interfaces. Sensors. September. 1988.

20. S. Sardella. New Computer Gives Disabled a "Voice". Boston Herald. November 20. 1994.

21. E. Schott. Uber die Registrierung des Nystagmus. Deutsches Archiv fur Klinische Medizin. 140 79-90. 1992.

22. E. Unger, M. Bijak, W. Mayr, C. Schmutterer and G. Schnetz. EOG-Controller for Rehabilitation Technology. The Fifth International Conference on Computers Helping People with Special Needs (ICCHP '96), Linz, Austria, July 1996. In J. Klaus; E. Auff; W. Kremser; W. Zagler (eds.) Interdisciplinary Aspects on Computers Helping People with Special Needs.. Vienna: R. Oldenbourg. 401 - 408. 1996.

23. H.A. Yanco. Wheelesley: a Robotic Wheelchair System; this volume. 1998.

24. H.A. Yanco and J. Gips. Preliminary Investigation of a Semi-Autonomous Robotic Wheelchair Directed Through Electrodes Proceedings of the Rehabilitation Engineering Society of North America Annual Conference, Pittsburgh, PA, RESNA Press, 1997, 414-416. 1997.

25. L.R. Young and D. Sheena. Survey of Eyemovement Recording Methods. Behavioral Research Methods and Instrumentation 7(5): 397-429. 1975.

Providing Intelligent Language Feedback for Augmentative Communication Users*

Christopher A. Pennington and Kathleen F. McCoy

Applied Science and Engineering Laboratories
Department of Computer and Information Sciences
University of Delaware
&
DuPont Hospital for Children
Wilmington, DE 19899
U.S.A.
{penningt,mccoy}@asel.udel.edu

Abstract. People with severe speech and motor impairments (SSMI) can often use augmentative communication devices to help them communicate. While these devices can provide speech synthesis or text output, the rate of communication is typically very slow. Consequently, augmentative communication users often develop telegraphic patterns of language usage. A natural language processing technique termed *compansion* (compression-expansion) has been developed that expands uninflected content words (i.e., compressed or telegraphic utterances) into syntactically and semantically well-formed sentences.

While originally designed as a rate enhancement technique, compansion may also be viewed as a potential tool to support English literacy for augmentative communication users. Accurate grammatical feedback from ill-formed inputs might be very beneficial in the learning process. However, the problems of dealing with inherently ambiguous errors and multiple corrections are not trivial. This paper proposes the addition of an adaptive user language model as a way to address some of these difficulties. It also discusses a possible implementation strategy using grammatical mal-rules for a prototype application that uses the compansion technique.

1 Introduction

People with severe speech and motor impairments (SSMI) can often use augmentative communication devices to help them communicate. While these devices

---

* This work has been supported by a Rehabilitation Engineering Research Center Grant from the National Institute on Disability and Rehabilitation Research of the U.S. Department of Education (#H133E30010) and by NSF Grant # IRI-9416916. Additional support has been provided by the Nemours Research Program.

V. O. Mittal et al. (Eds.): Assistive Technology and AI, LNAI 1458, pp. 59–72, 1998.
© Springer-Verlag Berlin Heidelberg 1998

can provide speech synthesis or text output, the rate of communication is typically very slow (most users average less than 10 words per minute). Consequently, augmentative communication users can often develop telegraphic patterns of language usage, especially if the disability occurs at an early age.

Although this functional style of communication is perfectly adequate for many situations, there are circumstances in which complete, grammatical English sentences are necessary to ensure proper communication and understanding. In addition, there are several obvious educational and psychological reasons for providing the ability to communicate in a literate manner. One in particular is to help dispel the general tendency of our society to automatically associate an inability to speak (or speak understandably) with a cognitive impairment or lack of intelligence.

To help address these concerns, a natural language processing technique termed *compansion* (compression-expansion) has been developed that expands uninflected content words (i.e., compressed or telegraphic utterances) into syntactically and semantically well-formed sentences [6], [15]. For example, given the input John go store yesterday, an intelligent augmentative communication system using compansion might produce "John went to the store yesterday."

Originally, compansion was designed as a rate enhancement technique for word- or symbol-based augmentative communication systems; that is, its primary purpose was to enable users to express themselves more quickly either using a speech synthesizer or for performing writing tasks. However, compansion can also be viewed as a potential tool to support English literacy efforts for augmentative communication users. This paper discusses the mechanisms needed to provide compansion with an enhanced ability to identify and correct language errors. A parallel effort for improving the written English of deaf people who are American Sign Language natives is also in progress [22,18].

2   Issues in Providing Intelligent Feedback

By providing accurate, grammatical feedback from ill-formed input, the compansion technique can be used to help facilitate the language development process, especially for users of symbol-based communication devices. At the very least, compansion can provide language reinforcement to the augmentative communication user through speech output and/or written text. This is analogous to the situation where a teacher or tutor would provide corrective instruction either verbally or visually (e.g., writing on a chalkboard).

Of course, there are several difficulties that must be dealt with to successfully provide *accurate* feedback. A basic issue is the ability to detect multiple errors in an ill-formed input. In addition, there may be potentially ambiguous interpretations of what those errors are, so properly identifying the errors is a major step. For example, John gone to the store could be incorrect because of a wrong past tense form ("John went to the store") or a missing auxiliary verb ("John had gone to the store").

Often, the combination of these factors will generate a whole set of possible corrections. Deciding which correction is the most appropriate can be very difficult. For example, The girl like John appears to have a subject-verb agreement error and could be corrected as "The girls like John" or "The girl likes John." However, for certain augmentative communication users, it could also be interpreted as "The girl was liked by John" or "The girls were liked by John." In some instances, the best suggestion for correction may be partially dependent on the specific user's language patterns.[1] The compansion technique already addresses these issues to some degree; nevertheless, there are several limitations that must be overcome in order to give truly intelligent feedback.

3   Overview of Compansion

The core of the compansion approach is a semantic parser that interprets input based on the use of case frames [8], [9]. Case frames are conceptual structures that represent the meaning of an utterance by describing the semantic cases or roles that each of the content words has in relationship with the others. In practice, the semantic parser designates the primary verb as the main component of the expression: all other words in the input are used to fill semantic roles with respect to the main verb that is chosen. While not an exhaustive or necessarily ideal list, we have adopted a set of semantic roles that have proven adequate for our purposes.

AGEXP (AGent/EXPeriencer) is the object doing the action, although for us the AGEXP does not necessarily imply intentionality, such as in predicate adjective sentences (e.g., *John* is the AGEXP in "John is happy"). THEME is the object being acted upon, while INSTR is the object or tool etc. used in performing the action of the verb. GOAL can be thought of as a receiver, and is not to be confused with BENEF, the beneficiary of the action. For example, in "John gave a book to Mary for Jane," *Mary* is the GOAL while *Jane* is the BENEF. We also have a LOC case which describes the event location (this case may be further decomposed into TO-LOC, FROM-LOC, and AT-LOC), and TIME which captures time and tense information (this case may also be subdivided).

As an example, given the input John go store, *go* would be selected as the main VERB, *John* would fill the role of AGEXP, and *store* would be assigned a LOC (LOCation) role.[2]

---

[1] Of course, the context in which the expression occurs is extremely important; however, in many cases it is not possible for an augmentative communication system to have access to both sides of the entire conversation, although advances in continuous speech recognition appear promising. At this point we only draw simple inferences from the partial context (e.g., tense changes); therefore, unless otherwise noted, utterances are considered in isolation.

[2] Note that it is ambiguous at this point whether *store* should be a TO-LOC or a FROM-LOC.

The semantic parser attempts to construct the most likely interpretation by filling in the various semantic roles in a way that makes the most sense semantically. It does this by employing a set of scoring heuristics which are based on the semantic types of the input words and a set of preferences which indicate which roles are the most important to be filled for a particular verb and what semantic types the fillers of the roles should have. The parser relies on a set of scoring heuristics (based on the preferences) to rate the possible interpretations (i.e., different ways of filling in the case frame) it comes up with [12]. "Idiosyncratic" case constraints specify which roles are mandatory or forbidden given a specific verb (or class of verbs). This captures, for example, the difference between transitive and intransitive verbs, where (in general) transitive verbs are required to have a theme, but a theme is forbidden with intransitive verbs. Other heuristics reflect general case preferences, including case importance (e.g., most verbs prefer THEMEs to be filled before BENEFiciaries), case filler (e.g., action verbs prefer animate AGEXPs), and case interactions (e.g., a human AGEXP might use an INSTRument, but an animal like a dog probably would not).

After all of the ratings for the various case preferences are assigned, they are combined together to produce a final score for each possible interpretation that the semantic parser produces. Any interpretation with a value less than a specified cut-off value is discarded, and the rest are ordered according to score and passed on for further processing. So, two possible parser interpretations of the input `apple eat John` might look like the following (DECL denotes a declarative expression):

```
(70 DECL
    (VERB (LEX EAT))
    (AGEXP (LEX JOHN))
    (THEME (LEX APPLE))
    (TENSE PRES))}
```

```
(20 DECL
    (VERB (LEX EAT))
    (AGEXP (LEX APPLE))
    (THEME (LEX JOHN))
    (TENSE PRES))
```

The first interpretation corresponds to a sentence such as "John eats the apple" while the second, lower rated interpretation corresponds to the interpretation "The apple eats John." Obviously, "John eats the apple" should be preferred over "The apple eats John" (in fact, the latter interpretation would almost always be discarded because of the previously mentioned cut-off value). The more likely interpretation is chosen because the preferences associated with the VERB *eat* strongly indicate a preference for an animate AGEXP (i.e., a human or animal) and a THEME that is edible.

Notice that the semantic reasoning that is the heart of compansion does not address syntactic realization issues. The semantic parser indicates that it prefers the interpretation where John is doing the eating and the `apple` is the thing being eaten. This preferred interpretation has many different syntactic realizations such as: "John eats the apple," "The apple was eaten by John," "It was the apple that John ate," etc.

4   Improving the Scoring Methodology

This rating system has proven useful for developing an initial research prototype of the compansion technique, allowing distinctions to be made about some important conceptual relationships. However, it must be improved upon if it is to be used to provide appropriate corrective feedback for augmentative communication users in the process of developing literacy skills. In addition, as pointed out above, these scores only capture preferences among semantic interpretations (e.g., "John likes pizza" vs. "John is liked by pizza") and provide little help with differentiating among several potential syntactic expressions of the same (or very similar) concept (e.g., "John likes pizza," "John likes the pizza," "The pizza is liked by John," etc.). This latter issue will be addressed in the next section.

Currently, most of the preference ratings for cases are based on intuition and the rules for combining scores are somewhat arbitrary. This is not sufficient to ensure a consistently reasonable set of possible corrections. Statistical data from tagged corpora should be used to provide better supported values for the ratings. Methods outlined in [1] suggest taking context into account as well as frequency when computing probabilities. A specific treatment of this approach for verb subcategorization is detailed in [23] and appears to be quite in line with our purposes. Information from lexical databases such as WordNet [19] is also being integrated to help improve part-of-speech and word sense preferences, as well as semantic classification information.

Furthermore, the functions used in combining scores should reflect an appropriate and well-established probabilistic method (see [3] for an overview of several possible algorithms). Related to this, the final scores should be normalized to provide a general measure of the appropriateness of an interpretation as well as to allow more objective comparisons between sentences.

Since the primary goal in this case is to promote literacy and not necessarily rate enhancement, a comprehensive list of choices should always be generated. This will increase the chances of augmentative communication users always finding a correct representation of what they want to express.[3] This does not detract, however, from the goal of presenting the best correction first whenever possible.

5   Accounting for Syntactic Variations

Besides the semantic parser, compansion also contains some rudimentary inferencing principles based on general observations of "telegraphic" forms of expression found in some sign languages and pidgins. For example, if no main verb is

---

[3] Of course there will always be instances in which compansion may be unable to correctly interpret the user's intended meaning. Even humans have a difficult time with that task from time to time.

found, it will attempt to infer the verbs *be* and/or *have*, taking into account the possible roles of the other words in the input. In a similar manner, if there is no valid agent, it will infer the pronouns *I* or *you*, depending on whether the input is a statement or a question. These techniques allow us to interpret input like happy ? to mean "Are you happy?" (as one reasonable possibility). At this point, we are beginning to reason about mainly syntactic distinctions and in fact, compansion uses a "word order" parser that attempts to account for various expected telegraphic and word order variations that may be common in this population (e.g., determining whether the output should maintain the input word order or not which would dictate whether the system should generate "John likes Mary" or "Mary is liked by John").

Additional research has begun that investigates more fully the often telegraphic language patterns of augmentative communication users [17]. Knowing more about general language expressions used in augmentative communication should enable compansion to make better choices among syntactic variants of the user's intended communication. The proposed methodology for accomplishing this is to group the common language variations into a taxonomy that can assist error identification [22].

Although there may be general language variations that occur, it is also likely that each individual will have idiosyncratic patterns of expression (e.g., some users may never construct passive sentences), including commonly made errors. This information could be very useful for error identification and for determining the most appropriate correction(s). Thus, there is a need for both an individual and a general user language model [4]. In addition, there is the possibility that an augmentative communication user's language abilities and preferences will change, especially if they are in the process of learning English literacy skills. This argues for a language model that can adapt to the user over time. This model will be essential for generating better interpretations, handling language errors intelligently, and providing additional feedback that may be helpful to the user.

6   An Adaptive User Language Model

In this section, we focus primarily on modeling syntactic expectations, given a specific user. We propose an adaptive user language model that requires several steps and relies on several different components to capture expectations for a particular user. First, a general language assessment model must be developed. This model will capture typical levels of literacy acquisition and indicate syntactic constructions that a person is expected to be able to use at a given level of acquisition. Second, this stereotypical model must be (possibly) modified for particular classes of users who may share common factors that could influence the process of language acquisition for that group (e.g., language transfer from a first language, prior literacy training, or speech therapy). The intermediate result of this design will be a model that captures expectations about syntactic structures being used by individuals that fall into various pre-defined levels of literacy expertise. The last component will be able to determine at what level to place a particular user, and to update the placement based on a carefully tracked history of interaction with that user. The final language model can then be used to help determine which suggested corrections are the most appropriate given the user's linguistic abilities and past language use.

6.1 SLALOM - A Language Assessment Model

Intuitively, people express themselves relatively consistently with a certain degree of language sophistication. Some grammar checkers rely on this concept when they assign a general "grade level" to a user's composition. Often this evaluation is based primarily on average sentence length (which is a very rough measure of the text's syntactic complexity). Knowing a person's writing "grade level" could help us immensely in choosing among various possible syntactic expressions. For instance, we would not expect someone who generally writes at a second grade level to use a complicated tense (e.g., past perfect) because that complexity is apparently beyond their current writing ability.

What we need is a mechanism that organizes syntactic constructions that are likely to be used together and can serve as the means for evaluating and predicting each user's English language proficiency. This "profile" can then be used to help determine a preferred interpretation when either the error or its underlying cause is ambiguous (e.g., when results from error identification suggest more than one possible correction for a single error).

To accomplish this, we propose the development of a language assessment model called SLALOM ("Steps of Language Acquisition in a Layered Organization Model") that is based on linguistic theories describing the processes by which people learn language. There is considerable linguistic evidence from research on both first language acquisition and second language acquisition that the acquisition order of language features is relatively consistent and fixed [11], [7], [2]. In fact, a stronger version of this statement is one of the central tenets of universal grammar theory (see for example, [10] and [13]).

The basic idea behind SLALOM is to divide the English language into a set of feature hierarchies (e.g., morphology, types of noun phrases, types of relative clauses) that are ordered from least to most complex. Features of similar complexity but in different hierarchies are arranged in layers that are then linked together to represent stereotypical "levels" of language ability.

Figure 1 contains an conceptual illustration of a piece of SLALOM.[4] We have depicted parts of four hierarchies in the figure: morphological syntactic features, noun phrases, verb complements, and various relative clauses. Within each hierarchy, the intention is to capture an ordering on the feature acquisition. So, for example, the model reflects the fact that the +ing progressive form of verbs is generally acquired before (and thus considered "easier to acquire" than)

---

[4] Specific details of the feature hierarchies have been simplified and are given for example purposes only.

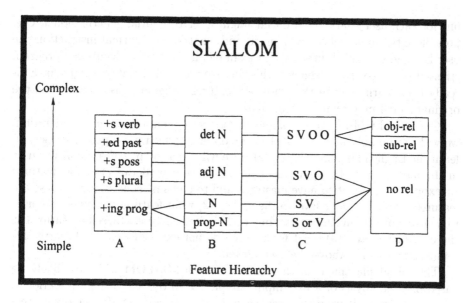

Fig. 1. Example of feature hierarchies in SLALOM

the +s plural form of nouns, which is generally acquired before the +s form of possessives, etc.

As mentioned before, the connections among the hierarchies are intended to capture sets of features which are acquired at approximately the same time. So, for example, the figure indicates that while the +s plural ending is being acquired, so too are both proper and regular nouns, and one and two word sentences. During this time, we do not expect to see any relative clauses.

These connections are derived from work in language assessment and grade expectations such as found in [14], and [5]. A combination of existing assessment tools will be needed to ensure adequate coverage of English language features and in some cases additional linguistics research may be required to develop an accurate and useful default model.

6.2 Customizing the Language Model

Once the default has been established, there must be a method for customizing the model on the basis of user characteristics which might cause either the simple/complex order within a hierarchy to change, or cause the predefined "levels" across the hierarchies to be altered. For instance, the model might be altered as a result of a study of texts from augmentative communication users. This study might result in an error taxonomy (mentioned earlier) which gives evidence (contrary to the views of universal grammar proponents) that the steps of language learning for augmentative communication users (as a group) were differently ordered from the standard expectations of English language acquisition. SLALOM can also be tailored to the needs of individual users via a series of "filters," one for each user characteristic that might alter the initial generic model. One possible filter could reflect how much and what kind of formal instruction the user has had in English literacy. For example, if the user's educational program stressed subject-verb agreement, this relatively complex feature might have already been learned, even though other features designated as "simpler" in the original model may remain problematic.

Once SLALOM has been completed for the population under consideration, we will presumably have a model of the order in which we expect our users to learn the English language. Essentially, we will need to "place" a particular user in the model. With this placement we will have a model of (1) what features we expect the student to have mastered and to be using consistently – these are features below the user's level in the model, (2) what features we expect the user to be using or attempting to use, but with limited success – these are features at the user's level, and (3) what features we do not expect to see used (correctly) – these are features above the user's level.

The initial placement of the student user in SLALOM will most likely be based on an analysis of a first input sample. Once this initial determination is made, further input from the user, as well as feedback given during any correction and tutorial phases, could cause the system to update the user's level in the model. It is important to note that although the default levels (i.e., cross-hierarchical connections) for the process of language acquisition will be somewhat predefined, the model is flexible enough to allow and account for individual variations beyond those represented by the initial model and its filters. In other words, additional information about each user's language usage gathered over time should provide a better and more accurate reflection of the current set of language features they are acquiring.

6.3 Adaptation Mechanisms

To realize a flexible model, a good history mechanism must be provided that can assist the language model in adapting to each individual's abilities and preferences. The history mechanism's responsibility is to update information in the user model based on experience with the augmentative communication user. Most of this information will be derived implicitly (e.g., analyzing expressive output to discover an especially problematic language feature), although a particular interface may allow explicit changes to the model.[5]

Potentially, there is a need for both a short-term and a long-term history mechanism. Short-term frequency data for errors and successes could be used to reassess the user's language abilities, especially when determining whether or not a specific language feature is known or in the process of being learned. This could be very helpful for deciding among several possible corrections as well as moving the user along the "steps" of the language model. Also, the prototypical language levels allow a system using this model to make reasonable

---

[5] This becomes more relevant if a tutoring component is being used to provide corrective responses.

default inferences when little knowledge is available. For example, if the user has not expressed a language feature before, the system can assume its acquisition level based on other features that are known.[6]

A long-term history mechanism would provide additional evidence for language change, as well as providing a way of adapting to the user's idiosyncratic language patterns. In addition, for tutorial purposes it might be useful to look for the user's avoidance of certain linguistic structures[7] since not all language difficulties are evident through error identification.

7   Developing a Compansion-Based Prototype

Up to this point, we have discussed two different aspects of a possible literacy aid for augmentative communication users: a semantic aspect based on conceptual information associated with words used in an input sentence, and a syntactic aspect based on expectations derived from the use of an adaptive language model. In this section, we discuss how these two aspects are being combined into an augmentative communication prototype.

The driving force of the process is a (primarily) syntactic grammar that is implemented in the Augmented Transition Network (ATN) formalism. An ATN parses sentences by encoding a grammar as a network of nodes and arcs (connecting the nodes) that is traversed as decisions are made about the input words (e.g., is it a noun, verb, etc.?). Registers containing specific information about the words and the parse may be passed through the network (and subnetworks), providing a powerful mechanism for reasoning about natural language. This formalism also allows arbitrary tests and actions to be performed during the parsing process; thus, many semantically-oriented tests can be incorporated directly into the grammar. Using this approach, we have encoded many aspects of the compansion technique into the grammar for this system; thus, the semantic "score" may be calculated as the grammar network is traversed. Below is a discussion of the changes needed to integrate an adaptive user language model into this application.

7.1   Using Mal-Rules to Encode Language Variations

The first step is to develop a syntactic grammar that is enhanced to capture the regular variants in the language use of augmentative communication users. A conceptual mechanism that could be used to simulate the language patterns would be mal-rules [21], [24]. Mal-rules are grammar rules specifically coded to accept ill-formed structures that reflect expected language errors; however, additional information can be associated with these rules that indicates an error has been made and what the possible cause(s) might be. The rules would handle observed telegraphic conventions (e.g., omitting forms of *be*) as well as any

---

[6] At this time it is not clear if the best strategy would be to assign the default as the minimum level, the highest level, or an average level.

[7] That we expect to see based on the perceived language level of the user.

commonly detected irregularities (e.g., inverted word order). A similar method has been used for second language learning [20].

A possible implementation of this approach is to construct a core grammar representing standard grammatical English and a separate set of mal-rules that captures common language variations of augmentative communication users. These mal-rules can be realized as an overlay of alternate arcs at the appropriate nodes within the ATN grammar. The resulting "modularity" will allow association of additional information with the mal-rules in a group-specific manner; for instance, we could construct appropriate error messages in this manner. If designed carefully, it should also be possible to (easily) use a different set of mal-rules (e.g., language patterns of a deaf person learning English as a second language) with the core grammar.

7.2   Implementing the User Language Model

In essence, this combination of mal-rules with the standard grammar comprises a "grammar" for all prototypical augmentative communication users. However, in order to individualize the grammar to specific users, a weighted grammar that assigns relative scores to each of its rules is proposed.[8] Usage frequency information from corpora and research data will be used as the initial weights for both the arcs of the standard grammar and the set of mal-rules. However, one complicating factor is that no large corpora exist for the language use of augmentative communication users. Thus, we must be careful in how the probabilities for the mal-rules are determined and rely mostly on data from standard text corpora.[9]

One possibility is that the initial values for the mal-rules will be predominantly stereotypical (i.e., reflecting the general relationships of the error taxonomy instead of being strictly frequency-based) and more sensitive to changes based on the user's interactions with the system. Some of the methods for dealing with sparse data [3] may also be helpful. In addition, features representing the relative complexity of acquisition will be attached to the nodes of the grammar. In the absence of other information, this value may be helpful in discriminating among multiple interpretations.

Once this default structure has been defined and initialized, the scores and features of the grammatical arcs (including those representing the mal-rules) may be modified by interactions with a separate user model that contains the individual's specific language characteristics (with respect to SLALOM). This model will consist of long-term information including the following: what language features are known, unknown, or in the process of acquisition; an overall measure of the user's language level (derived from the known language features);

---

[8] The most likely implementation is a probabilistic context-free grammar similar to those described by Charniak [3] and Allen [1].

[9] In a related effort, we are engaged in an ongoing process of collecting conversational data (keystrokes, "spoken" text and some video) from a few augmentative communication users and hope to take advantage of this information at some point. It is unclear if this will be generalizable, though.

and historical data reflecting the user's language usage and error patterns. The latter information will be used to make changes to the grammar for each particular user.

Eventually, these changes will allow the grammar to adapt to the augmentative communication user's specific language style. Exact criteria for deciding when to change the feature acquisition values (e.g., from "acquiring" to "known") have not yet been determined, but essentially we can view the placement in SLALOM as highlighting language features (and corresponding mal-rules) that we expect the user to be learning at a given point in time. Thus, it gives us a glimpse into users' current language patterns by zeroing in on the mal-rules we expect them to be using at this point in their acquisition of English. Key to this process is the feedback provided by interactions where one of the suggested corrections is selected. This information will help to either confirm or modify the system's current "view" of the user. In any event, the mechanisms needed to implement these adjustments should be straightforward.

7.3   Processing Considerations

After a sentence is parsed, the identified errors will be returned and tagged appropriately with the mal-rule(s) thought to be responsible. In many cases, we cannot assume that there will be a one-to-one mapping between the identified mal-rules and the possible corrections. Confounding this issue is the strong possibility of multiple errors in each sentence, possibly interacting with each other; hence, it might be necessary to look at evaluating sets of mal-rules that are triggered instead of individual ones. At this time it is unclear what method will be best for determining the most likely set of mal-rules.

8   Future Work

The most immediate need is to further specify the relationships of features within SLALOM and their likeliness to occur. In addition, while there is some evidence of what constitutes a "typical" telegraphic language pattern, more work must be done to classify these variations and to gain information on their frequency of use. Once this is accomplished, the data can be used in the modifications that will be made to the current compansion-based application as it integrates the adaptive language model. As discussed previously, it is thought that these changes will take the form of adding mal-rules and weighted features to the ATN, along with any necessary reasoning mechanisms. Adaptability will be addressed by superimposing a history mechanism that will adjust weights and other features based on experiences with the augmentative communication user's language choices and feedback selections.

Results from this work will be filtered back into a larger project called ICICLE (Interactive Computer Identification and Correction of Language Errors). ICICLE currently encompasses the mechanisms for identifying errors in the written English of deaf people. As mentioned earlier, the design of corrective feedback mechanisms for that system is proceeding in parallel with the work described here. It is hoped that some of the semantic reasoning strategies in compansion will be of use to ICICLE as well.

Another essential component being designed for ICICLE concerns adaptive tutoring and explanation [16]. This module will be able to consult the adaptive language model for information to help customize its instruction for the individual user. Finally, at the present time, both ICICLE and the compansion technique are primarily concerned with clause- or sentence-level variations; however, it is important to note that many difficulties in English literacy occur at a discourse level (e.g., anaphora resolution). This is a major area of needed research.

9 Summary

The compansion technique has great potential for use as a tool to help promote literacy among users of augmentative communication systems. By providing linguistically correct interpretations of ill-formed input, it can reinforce proper language constructions for augmentative communication users who are in the process of learning English or who have developed telegraphic patterns of language usage. To accomplish this goal, several modifications to the existing compansion approach are proposed to improve the accuracy of the corrective feedback. The most significant change is the addition of an adaptive language model. This model initially provides principled defaults that can be used to help guide the identification and correction of language errors, adapting to each user's specific language abilities and patterns over time. Finally there is a discussion of using sets of grammatical mal-rules to integrate the language model into an existing application that uses the compansion technique.

References

1. J. Allen. *Natural Language Understanding.* Benjamin/Cummings, Redwood City, CA, 1995.
2. N. Bailey, C. Madden, and S. D. Krashen. Is there a "natural sequence" in adult second language processing? *Language Learning*, 24(2):235–243, 1974.
3. E. Charniak. *Statistical Language Learning.* MIT Press, Cambridge, MA, 1993.
4. D. N. Chin. KNOME: Modeling what the user knows in UC. In Alfred Kobsa and Wolfgang Wahlster, editors, *User Models in Dialog Systems*, Berlin Heidelberg New York Tokyo, 1989. Springer.
5. D. Crystal. *Profiling Linguistic Disability.* Edward Arnold, London, 1982.
6. P. W. Demasco and K. F. McCoy. Generating text from compressed input: An intelligent interface for people with severe motor impairments. *Communications of the ACM*, 35(5):68–78, May 1992.
7. H. C. Dulay and M. K. Burt. Natural sequences in child second language acquisition. *Language Learning*, 24(1):37–53, 1974.
8. C. J. Fillmore. The case for case. In E. Bach and R. Harms, editors, *Universals in Linguistic Theory*, pages 1–90, New York, 1968. Holt, Rinehart, and Winston.

9. C. J. Fillmore. The case for case reopened. In P. Cole and J. M. Sadock, editors, *Syntax and Semantics VIII: Grammatical Relations*, pages 59–81, New York, 1977. Academic Press.

10. J. A. Hawkins. Language universals in relation to acquisition and change: A tribute to Roman Jakobson. In Linda R. Waugh and Stephen Rudy, editors, *New Vistas in Grammar: Invariance and Variation*, pages 473–493. John Benjamins, Amsterdam / Philadelphia, 1991.

11. D. Ingram. *First Language Acquisition: Method, Description, and Explanation.* Cambridge University Press, Cambridge; New York, 1989.

12. M. Jones, P. Demasco, K. McCoy, and C. Pennington. Knowledge representation considerations for a domain independent semantic parser. In J. J. Presperin, editor, *Proceedings of the Fourteenth Annual RESNA Conference*, pages 109–111, Washington, D.C., 1991. RESNA Press.

13. E. L. Keenan and S. Hawkins. The psychological validity of the accessibility hierarchy. In Edward L. Keenan, editor, *Universal Grammar: 15 Essays*, pages 60–85. Croon Helm, London, 1987.

14. L. L. Lee. *Developmental Sentence Analysis: A Grammatical Assessment Procedure for Speech and Language Clinicians.* Northwestern University Press, Evanston, IL, 1974.

15. K. F. McCoy, P. W. Demasco, M. A. Jones, C. A. Pennington, P. B. Vanderheyden, and W. M. Zickus. A communication tool for people with disabilities: Lexical semantics for filling in the pieces. In *Proceedings of the First Annual ACM Conference on Assistive Technologies*, pages 107–114, New York, 1994. ACM.

16. K. F. McCoy and L. N. Masterman. A tutor for deaf users of American sign language. In *Proceedings of Natural Language Processing for Communication Aids, an ACL/EACL '97 Workshop*, Madrid, Spain, July 1997.

17. K. F. McCoy, W. M. McKnitt, C. A. Pennington, D. M. Peischl, P. B. Vanderheyden, and P. W. Demasco. AAC-user therapist interactions: Preliminary linguistic observations and implications for Compansion. In Mary Binion, editor, *Proceedings of the RESNA '94 Annual Conference*, pages 129–131, Arlington, VA, 1994. RESNA Press.

18. K. F. McCoy, C. A. Pennington, and L. Z. Suri. English error correction: A syntactic user model based on principled "mal-rule" scoring. In *Proceedings of UM-96, the Fifth International Conference on User-Modeling*, pages 59–66, Kailua-Kona, HI, January 1996.

19. G. A. Miller. Wordnet: A lexical database for English. *Communications of the ACM*, pages 39–41, November 1995.

20. E. Schuster. Grammars as user models. In *Proceedings of IJCAI 85*, 1985.

21. D. H. Sleeman. Inferring (mal) rules from pupil's protocols. In *Proceedings of ECAL-82*, pages 160–164, Lisay, France, 1982.

22. L. Z. Suri and K. F. McCoy. Correcting discourse-level errors in CALL systems for second language learners. *Computer-Assisted Language Learning*, 6(3):215–231, 1993.

23. A. Ushioda, D. A. Evans, T. Gibson, and A. Waibel. Frequency estimation of verb subcategorization frames based on syntactic and multidimensional statistical analysis. In *Proceedings of the 3rd International Workshop on Parsing Technologies (IWPT3)*, Tilburg, The Netherlands, August 1993.

24. R. M. Weischedel and N. K. Sondheimer. Meta-rules as a basis for processing ill-formed input. *American Journal of Computational Linguistics*, 9(3-4):161–177, July-December 1983.

Saliency in Human-Computer Interaction

Polly K. Pook*

IS Robotics, Inc.
22 McGrath Highway
Somerville, MA 02143
U.S.A.
pook@isr.com

Abstract. This paper considers ways in which a person can cue and constrain an artificial agent's attention to salient features. In one experiment, a person uses gestures to direct an otherwise autonomous robot hand through a known task. Each gesture instantiates the key spatial and intentional features for the task at that moment in time. In a second hypothetical task, a person uses speech and gesture to assist an "intelligent room" in learning to recognize the objects in its environment. Both experiments use a system of dynamic reference, termed *deictic*, to bind the robot's perception and action to salient task variables.

1 Introduction

In assistive technology, a person directs a robotic device with varying degrees of control. For example, one desires executive control over an intelligent wheelchair. We want to be able to say "Go there" without bothering with the details of navigating around obstacles or accounting for particular door handle shapes. Alternatively, one often needs finely tuned control over a prosthetic device in order to perform new and/or arbitrary tasks, such as picking up a new instrument or, as a friend demonstrated recently, grabbing the rope on a windsurfer sail. The appropriate level of control, however, is not necessarily specific to the device. There are times when the specific motion of a wheelchair should be under the person's guidance. Similarly, advances in autonomous prosthetics, to control balance, dynamics and reflexive positioning without conscious monitoring, are welcome. Research in assistive technology must cover the spectrum of control interfaces.

This paper considers two examples of human-robot interaction within the narrow region of low-level robotic control. In one, a person assists a robot manipulator to blindly open a door. This experiment considers what key information can be communicated to the robot using a simple gestural sign language. The

---

* The work on *teleassistance* was conducted at the University of Rochester and supported by NSF and the Human Science Frontiers Program. Research in the MIT Intelligent Room was supported by DARPA.

V. O. Mittal et al. (Eds.): Assistive Technology and AI, LNAI 1458, pp. 73–83, 1998.
© Springer-Verlag Berlin Heidelberg 1998

second example is a hypothetical situation in which a person helps a "robotic room" to find an object in its visual scene. This considers the use of context-dependent speech cues. Note that the motivation behind the two examples is the opposite of the workshop precis: we look not at what's helpful to a person, but rather at what's helpful to the robot.

Why consider the robot's needs rather than the person's? One reason is self-motivated: I'm a robotic researcher not a medical assistant. I naturally find myself, as do many roboticists I suspect, endeavoring to assist the robot perform and learn. Caveat emptor.

A better reason is that autonomous robots are limited to the current understanding and implementation of artificial intelligence. An immediate way to extend the robot's capabilities is to add on-line human intelligence, to "put the human in the loop."

But how? What knowledge does the robot need from the person and how does the person communicate that information economically? These questions drive the third rationale for examining human-robot interaction from the robot's side. Consider modelling the human-robot paradigm as a single entity, with the robot performing perceptuo-motor reflex behaviors and the person performing high-level cognitive processing. The bandwidth between the two must necessarily be low, forcing cognitive input to be salient to the task at hand. By starting with the robot's needs, we examine the question of what is salient from the bottom-up.

2 A Deictic Strategy

We use the term *deictic*, from the Greek *deiktikos* meaning pointing or showing, to describe the mechanism by which the person interacts with the robot in the two examples presented in this paper.

Linguists classify the words *this* and *that* as deictic because they "point" to a referent in the world (be it physical or virtual). The pointer suffices to bind the listener's interpretation to the referent, without requiring a unique identification. A key advantage of deixis is its dynamicism. "This book" constrains reasoning about the utterances immediately preceding and succeeding the phrase. The listener is bound to a context, defined by "this book," and is able to interpret surrounding statements accordingly. Hearing "that book" or "this cup" immediately shifts the binding to a new context. The listener doesn't need to keep the old context around (unless that book is about this cup, but such recursion is limited). Consequently, momentary deictic binding allows the listener to constrain interpretation to a particular context and optimize the use of short-term memory.

2.1 Deictic Vision

Psychophysical studies suggest that animals use deictic strategies to bind cognitive goals to spatial targets and low-level perceptual and motor programs [8,4].

Visual fixation is another example of a deictic strategy. The act of fixating on an object centers the target in the retinotopic array, potentially simplifying cognitive processes to deal with "the object I'm looking at" rather than with the attributes that distinguish that particular object from all others [12,7]. Moreover, spatial relationships can be defined relative to the object and so be invariant to the viewer's location in world coordinates.

The perceptual feedback required by the cognitive process is constrained by the process itself. Ballard and Hayhoe find that saccadic eye movements appear to correspond to minimal, sequential visual processes[4,3]. In a task in which subjects were asked to arrange a set of colored blocks to match a model, the subjects most often performed two saccades for each block. Presumably, one fixation acquires the color of the target block and another acquire its relative location. They posit that only color is actively interpreted in the first case; position in the second. By this theory, the sensory interpretation is bound to working variables in memory to support the current process. These are wiped clean by the next deictically bound process. This method would drastically simplify computational requirements to the specific variables, or degrees of freedom, of the current decision process.

Research in real-time computer vision has built on the deictic theory. Whitehead and Ballard [13] model visual attention as a binding strategy that instantiates a small set of variables in a "blocks-world" task. The agent's "attention" binds the goal to a visual target. Success in computer vision stems predominantly from special purpose processes that examine only a subset of the sensorium: for example, optic flow for motion detection, color histogramming for object recognition and zero-disparity filters for tracking.

2.2   Deictic Reference in Robots

In robotics, deictic variables can define relative coordinate frames for successive motor behaviors [1]. Such variables can avoid world-centered geometry that varies with robot movement. To open a door, for instance, looking at the doorknob defines a relative servo target. In another chapter in this book, Crisman and Cleary [6] discuss the computational advantage of target-centered reference frames for mobile robot navigation. Hand and body position also provide a relative reference frame. Since morphology determines much of how hands are used, the domain knowledge inherent in the shape and frame position can be exploited. The features salient to the task (direction, force) can be extracted and interpreted within the constraints of the reference frame.

3   Example 1: Gestures for Robot Control

Understanding the deictic references used to bind cognitive programs gives us a starting model for human/robot interaction. In this model, which we call *tele-assistance* [10] the human provides the references via hand gestures and an otherwise autonomous robot carries out the motor programs. A gesture selects the next motor program to perform and tunes it with hand-centered markers. This illustrates a way of decoupling the human's link between motor program and reflexes.

The dual-control strategy of teleassistance combines teleoperation and autonomous servo control to their advantage. The use of a simple sign language abstracts away many problems inherent to literal master/slave teleoperation. Conversely, the integration of global operator guidance and hand-centered coordinate frames permits the servo routines to position the robot in relative coordinates and perceive features in the feedback within a constrained context, significantly simplifying the computation and reducing the need for detailed task models.

In these experiments the human operator wears a data glove (an EXOS hand master) to communicate the gestures, such as pointing to objects and adopting a grasp preshape. Each sign indicates intention: e.g., reaching or grasping; and, where applicable, a spatial context: e.g., the pointing axis or preshape frame. The robot, a Utah/MIT hand on a Puma arm, acts under local servo control within the proscribed contexts.

3.1   Opening a Door

The gestural language is very simple. To assist a robot to open a door requires only three signs: point, preshape, and halt. Pointing to the door handle prompts the robot to reach toward it and provides the axis along which to reach. Preshaping indicates the handle type. The halting sign stops all robot motion and is included only as a precautionary measure. Pointing and preshaping are purposefully imprecise; the reactive servo routines on the robot provide fine control locally.

A finite state machine (FSM) for the task specifies the flow of control (Figure 1). This embeds gesture recognition and motor response within the overall task context.

Pointing and preshaping the hand create hand-centered spatial frames. Pointing defines a relative axis for subsequent motion. In the case of preshaping, the relative frame attaches within the opposition space [2] of the robot fingers. For example, a wrap grasp defines a coordinate system relative to the palm. With adequate dexterity and compliance, simply flexing the robot fingers toward the origin of the frame coupled with a force control loop suffices to form a stable grasp. Since the motor action is bound to the local context, the same grasping action can be applied to different objects – a spatula, a mug, a doorknob – by changing the preshape.

The two salient features for each motor program in this task are direction, specified by the hand signs, and force, specified as a significant change in tension in any of the finger joints. Force is perceived identically but interpreted differently according to the motor program. While reaching, force suggests that the door has been reached, while turning a force is interpreted as contact with the mechanical stop, i.e., the knob is fully turned. The bound context permits the program to

FSM for "Open a Door"

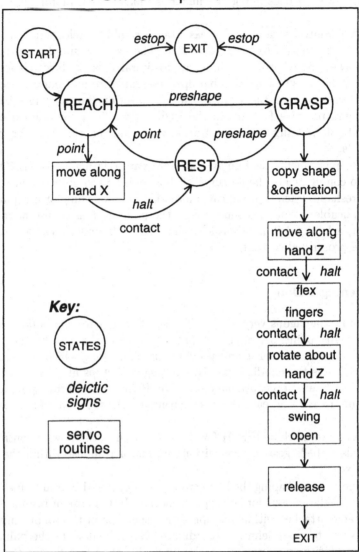

Fig. 1. A finite state machine (FSM) for opening a door. The operator's hand sign causes a transition to the appropriate state and, from there, to the corresponding servo routines. The state determines the motor program to execute. The deictic hand signs, *point* and *preshape*, provide a spatial coordinate frame. The routines servo on qualitative changes in joint position error that signify a force contact. Each contact is interpreted within the current context, e.g., bumping the door or reaching the knob's mechanical stop. No special force model is needed. The operator can also push the flow of control through the servo routines by signaling the *halt* hand sign.

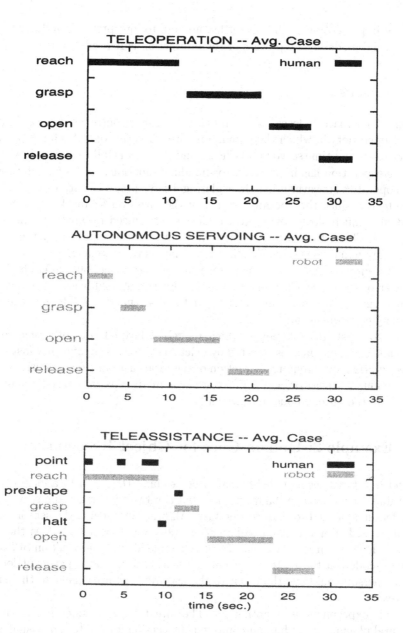

Fig. 2. Plots of the time that the human operator (dark bars) and the autonomous robot routines (grey bars) actively control the robot during each phase of the task, under the three control strategies. The teleoperator (top) must supervise 100% of the task; under autonomous control (middle), the robot is fully in charge but with limited strategic abilities; in teleassistance (bottom) the operator supervises the robot only 25% of the time for this task. Once the hand is teleassisted to a position near the door handle, the robot completes the task autonomously.

constrain perception to the salient feature and to interpret it in a dynamic way. No special model of the physics of the interaction is needed.

3.2 Results

Figure 2 compares teleassistance to two other robot control strategies. The first is teleoperation, in which the human operator directly controls all robot motion in a closed, real-time servo loop. The second strategy is fully autonomous robots.

Teleoperation has improved error-handling functions, as supplied by the human operator. However it has three glaring disadvantages. First, it requires 100% monitoring, since the operator is part of the low-level feedback loops. Second, control is much slower owing to the delays introduced by putting a human in the loop. Third, the robot is vulnerable to inaccurate responses by the human, which are numerous under significant communication latencies.

The autonomous strategy is faster than teleassistance, but suffers by having little error-tolerance. Current real-time robots cannot readily accommodate even simple changes in their environment. If the door handle is different the door-opening experiment fails.

In contrast, teleassistance, which models a layered control structure that uses autonomous routines directed by deictic gestures, is significantly faster than teleoperation, only requires a fraction of the total task time for executive control by the human operator, and can better accommodate natural variations in tasks than can the autonomous routines alone.

4   Example 2: Human Cues for Object Recognition

In teleassistance, economical human gestures bind the robot to a local context. Within that momentary binding, the robot can readily extract and servo on the key features needed to perform the task. All other features in the sensorium can be disregarded or relegated to lower-level controllers. But how does the robot know which features are salient? In the example above, the selection of features is hard-coded in the motor program. This next, proposed experiment looks at how a person could cue the robot to the features that are salient to the current context.

This experiment is hypothetical. The intent is to consider linguistic and gestural phenomena in human-computer interaction for the task of visual object recognition. Natural human cues can advantageously constrain computation as the computer learns to identify, label, and later recognize and locate objects in a room. The method would exploit spoken phrases by associating syntactic and semantic features with visual features of the object and its surroundings; constrain the visual attention to the area pointed to; and extract additional delimiting features from iconic gestures. The project addresses issues in linguistics, human-computer interaction, machine learning, computer vision, and gesture understanding.

The platform is the MIT AI Lab Intelligent Room (for details see [11,5], or the web site at http://www.ai.mit.edu/projects/hci). The room is an ordinary conference room equipped with multiple cameras, microphones, various displays, a speech synthesizer, and nine dedicated workstations. The room is able to be aware of people, by means of visual tracking and rudimentary gesture recognition, and to be commanded through speech, keyboard, pointing, or mouse.

4.1   Background

To recognize an object, computer vision algorithms first analyze features in an image or set of camera views. Which features are salient depends in part on the object. The perception of motion, for instance, doesn't help one identify an apple but is often used to locate people in a scene. Saliency can be heavily dependent on context. For example, color perception is useful when searching for a particular black chair among a roomful of red ones; it is not helpful when all the chairs are black. Note that salience means *noticeable*, not *intrinsic* (although an intrinsic property may well be a salient one).

4.2   Human Supervision

Research in computer vision has been successful in designing algorithms that recognize particular features, such as color or edges. Success is bounded, however, by an inability to dynamically select which features are relevant given the target object and context. One can explore this boundary by including human supervisory input during the recognition process. A person can prioritize the feature search by means of natural speech and gesture. Consider a sample scenario. A person points and asks the room "What is that?" The room agent scans the scene in the indicated direction and looks for a predefined set of features. If the computer is unable to detect an object or selects the wrong one, the person can correct the search strategy by highlighting the salient features verbally, for example by saying "no, it's black."

The interaction establishes constraints on the computational process in several ways. By pointing, the person constrains the search area; the speech syntax delimits an initial feature ordering; and the semantic content highlights key features and provides an error signal on the correctness of the match. Within the momentary constraint system defined by the interaction, the appropriate feature detection algorithms can be applied selectively for the task of object recognition.

We will use this system to extend information query to include a 3D interface. Previously, query systems relied on keyboard input solely. More recently, there has been work in integrating 2D spatial data in the query, such as by selecting coordinates on a map with a mouse or pointer. In this proposal, one can extend the interface to include queries about physical objects in a room. For example one might point to a VCR and ask what it is and how to operate it.

Sample Object Recognition Scenario

1. LEARNING

Person
  Speech: "That is a VCR"
  Gesture: Pointing

Computation
  Speech analysis:
    Analyze syntactic structure
    Search for known object "VCR"
    Store discourse context
  Gesture analysis:
    Circumscribe visual attention to the pointing direction
  Visual processing:
    Initialize visual feature map with known or discerned
     parameters
    Visual search within attention cone

Computer Output
  Speech: Affirm or request clarification
  Gesture: Highlight candidate object if found

2. CORRECTION Person
  Speech: "No, it's black" (or "bigger", "to the left", etc.)
  Gesture: Pointing or Iconic (e.g., describes object shape
    or size)

Computation
  Speech analysis:
    Analyze syntactic structure
    Refer to discourse context
  Gesture analysis:
    Adjust attention to new direction
    Recognize iconic gesture and extract relevant spatial
     parameters
  Visual processing:
    Tune visual feature map with new parameters
    Repeat visual search

Computer Output
  Speech: Affirm or request clarification
  Gesture: Highlight candidate object

3. LEARNED

Person
  Speech: "Right"

Computation
  Extract other camera views
  Store salient features
  Label room contents database

```
Computer Output
  Spoken: affirm
```

4. RECALL
Person
```
  Speech: "What is that?" "Where is the VCR?"
  Gesture: Pointing
```
Computation
```
  Search labeled database for feature map
  Visual search
  Select instance nearest to the person
```
Computer Output
```
  Spoken: affirm or request help
  Gesture: Highlight candidate object
```

5. INFORMATION QUERY
Person
```
  Speech: "How does this work?"
  Gesture: Pointing
```
Computation
```
  Match recognized image against database template.
```
Computer Output
```
  Visual and verbal instructions on use.
```

5   Conclusion

Determining saliency is a significant problem in AI. To us, what's so striking about a visual scene or a physical force is, to an artificial agent, often indistinguishable from the rest of the sensorium. The difficulty is compounded by context dependency: features that predominate in one setting are often irrelevant to another. The agent therefore could benefit greatly from on-line human cues.

There is obvious application for human supervision in assistive technology. Less obviously, it allows researchers to consider the integration of existing technologies in more interesting and complex domains than would be permitted if the computer had to operate autonomously. For example, one can study the influence of spoken cues on computer vision without a working model of the cognitive structure that prompts the speech. What visual, oral, and gestural features are important when? What are the essential cues needed by reactive artificial agents? These questions go toward designing communication interfaces for both the disabled and the able-bodied.

The work proposed here limits consideration to features that are rather low-level: color, shape, direction, etc. The research may have application to areas such as prosthetics, but it is worth noting that assistive technology certainly seeks higher-level control as well.

References

1. P. E. Agre and D. Chapman. Pengi: an implementation of a theory of activity. In *Proceedings of the Sixth National Conference on Artificial Intelligence*, pages 268–272. Morgan Kaufmann, Los Altos, CA, 1987.
2. M. Arbib, T. Iberall, and D. Lyons. Coordinated control programs for movements of the hand. Technical report, COINS Department of Computer and Information Science, University of Massachussetts, 1985.
3. D. H. Ballard, M. M. Hayhoe, P. K. Pook, and R. Rao. Deictic codes for the embodiment of cognition. *The Behavioral and Brain Sciences*, 1997. [To appear – earlier version available as National Resource Laboratory for the study of Brain and Behavior TR95.1, January 1995, U. of Rochester].
4. D.H. Ballard, M.M. Hayhoe, F. Li, and S.D. Whitehead. Hand-eye coordination during sequential tasks. *Phil. Transactions of the Royal Society of London*, March 1992.
5. M.H. Coen. Building brains for rooms: Designing distributed software agents. In *Proceedings of the AAAI National Conference*, August 1997.
6. J.D. Crisman and M.E. Cleary. Progress on the deictically controlled wheelchair. In Mittal et al. [9]. This volume.
7. J.J. Gibson. *The Perception of the Visual World*. Houghton Mifflin, Boston, 1950.
8. E. Kowler and S. Anton. Reading twisted text: Implications for the role of saccades. *Vision Research*, pages 27:45–60, 1987.
9. V. Mittal, H.A. Yanco, J. Aronis and R. Simpson, eds. *Lecture Notes in Artificial Intelligence: Assistive Technology and Artificial Intelligence*. Springer-Verlag, 1998. This volume.
10. P. K. Pook and D. H. Ballard. Deictic teleassistance. In *Proceedings of the IEEE/RSJ/GI International Conference on Intelligent Robots and Systems (IROS)*, September 1994.
11. M.C. Torrance. Advances in human-computer interaction: The intelligent room. In *Research Symposium, Human Factors in Computing: CHI'95 Conference*, May 1995.
12. S. Ullman. Against direct perception. *The Behavioral and Brain Sciences*, 3:373–415, 1980.
13. S. D. Whitehead and D. H. Ballard. A preliminary study of cooperative mechanisms for faster reinforcement learning. University of Rochester, Department of Computer Science, November 1990.

A Wearable Computer Based American Sign Language Recognizer

Thad Starner, Joshua Weaver, and Alex Pentland

The Media Laboratory
Massachusetts Institute of Technology
20 Ames Street
Cambridge MA 02139
{thad,joshw,sandy}@media.mit.edu

Abstract. Modern wearable computer designs package workstation level performance in systems small enough to be worn as clothing. These machines enable technology to be brought where it is needed the most for the handicapped: everyday mobile environments. This paper describes a research effort to make a wearable computer that can recognize (with the possible goal of translating) sentence level American Sign Language (ASL) using only a baseball cap mounted camera for input. Current accuracy exceeds 97% per word on a 40 word lexicon.

1 Introduction

While there are many different types of gestures, the most structured sets belong to the sign languages. In sign language, where each gesture already has assigned meaning, strong rules of context and grammar may be applied to make recognition tractable.

To date, most work on sign language recognition has employed expensive "datagloves" which tether the user to a stationary machine [26] or computer vision systems limited to a calibrated area [23]. In addition, these systems have mostly concentrated on finger spelling, in which the user signs each word with finger and hand positions corresponding to the letters of the alphabet [6]. However, most signing does not involve finger spelling, but instead uses gestures which represent whole words, allowing signed conversations to proceed at or above the pace of spoken conversation.

In this paper, we describe an extensible system which uses one color camera pointed down from the brim of a baseball cap to track the wearer's hands in real time and interpret American Sign Language (ASL) using Hidden Markov Models (HMM's). The computation environment is being prototyped on a SGI Indy; however, the target platform is a self-contained 586 wearable computer with DSP co-processor. The eventual goal is a system that can translate the wearer's sign language into spoken English. The hand tracking stage of the system does not attempt a fine description of hand shape; studies of human sign

V. O. Mittal et al. (Eds.): Assistive Technology and AI, LNAI 1458, pp. 84–96, 1998.
© Springer-Verlag Berlin Heidelberg 1998

readers have shown that such detailed information is not necessary for humans to interpret sign language [18,22]. Instead, the tracking process produces only a coarse description of hand shape, orientation, and trajectory. The hands are tracked by their color: in the first experiment via solidly colored gloves and in the second, via their natural skin tone. In both cases the resultant shape, orientation, and trajectory information is input to a HMM for recognition of the signed words.

Hidden Markov models have intrinsic properties which make them very attractive for sign language recognition. Explicit segmentation on the word level is not necessary for either training or recognition [25]. Language and context models can be applied on several different levels, and much related development of this technology has already been done by the speech recognition community [9]. Consequently, sign language recognition seems an ideal machine vision application of HMM technology, offering the benefits of problem scalability, well defined meanings, a pre-determined language model, a large base of users, and immediate applications for a recognizer.

American Sign Language (ASL) is the language of choice for most deaf people in the United States. ASL's grammar allows more flexibility in word order than English and sometimes uses redundancy for emphasis. Another variant, Signing Exact English (SEE), has more in common with spoken English but is not in widespread use in America. ASL uses approximately 6000 gestures for common words and communicates obscure words or proper nouns through finger spelling.

Conversants in ASL may describe a person, place, or thing and then point to a place in space to store that object temporarily for later reference [22]. For the purposes of this experiment, this aspect of ASL will be ignored. Furthermore, in ASL the eyebrows are raised for a question, relaxed for a statement, and furrowed for a directive. While we have also built systems that track facial features [7], this source of information will not be used to aid recognition in the task addressed here.

The scope of this work is not to create a user independent, full lexicon system for recognizing ASL, but the system should be extensible toward this goal. Another goal is real-time recognition which allows easier experimentation, demonstrates the possibility of a commercial product in the future, and simplifies archiving of test data. "Continuous" sign language recognition of full sentences is necessary to demonstrate the feasibility of recognizing complicated series of gestures. Of course, a low error rate is also a high priority. For this recognition system, sentences of the form "personal pronoun, verb, noun, adjective, (the same) personal pronoun" are to be recognized. This sentence structure emphasizes the need for a distinct grammar for ASL recognition and allows a large variety of meaningful sentences to be generated randomly using words from each class. Table 1 shows the words chosen for each class. Six personal pronouns, nine verbs, twenty nouns, and five adjectives are included making a total lexicon of forty words. The words were chosen by paging through Humphries et al. [10] and selecting those which would generate coherent sentences when chosen randomly for each part of speech.

Table 1. ASL Test Lexicon

part of speech	vocabulary
pronoun	I, you, he, we, you(pl), they
verb	want, like, lose, dontwant, dontlike, love, pack, hit, loan
noun	box, car, book, table, paper, pants, bicycle, bottle, can, wristwatch, umbrella, coat, pencil, shoes, food, magazine, fish, mouse, pill, bowl
adjective	red, brown, black, gray, yellow

Attempts at machine sign language recognition have begun to appear in the literature over the past five years. However, these systems have generally concentrated on isolated signs, immobile systems, and small training and test sets. Research in the area can be divided into image based systems and instrumented glove systems.

Tamura and Kawasaki demonstrate an early image processing system which recognizes 20 Japanese signs based on matching cheremes [27]. Charayaphan and Marble [3] demonstrate a feature set that distinguishes between the 31 isolated ASL signs in their training set (which also acts as the test set). More recently, Cui and Weng [4] have shown an image-based system with 96% accuracy on 28 isolated gestures.

Takahashi and Kishino [26] discuss a user dependent Dataglove-based system that recognizes 34 of the 46 Japanese kana alphabet gestures, isolated in time, using a joint angle and hand orientation coding technique. Murakami and Taguchi [17] describe a similar Dataglove system using recurrent neural networks. However, in this experiment a 42 static-pose finger alphabet is used, and the system achieves up to 98% recognition for trainers of the system and 77% for users not in the training set. This study also demonstrates a separate 10 word gesture lexicon with user dependent accuracies up to 96% in constrained situations. With minimal training, the glove system discussed by Lee and Xu [13] can recognize 14 isolated finger signs using a HMM representation. Messing et. al. [16] have shown a neural net based glove system that recognizes isolated fingerspelling with 96.5% accuracy after 30 training samples. Kadous [12] describes an inexpensive glove-based system using instance-based learning which can recognize 95 discrete Auslan (Australian Sign Language) signs with 80% accuracy. However, the most encouraging work with glove-based recognizers comes from Liang and Ouhyoung's recent treatment of Taiwanese Sign language [14].

This HMM-based system recognizes 51 postures, 8 orientations, and 8 motion primitives. When combined, these constituents form a lexicon of 250 words which can be continuously recognized in real-time with 90.5% accuracy.

2   Use of Hidden Markov Models in Gesture Recognition

While the continuous speech recognition community adopted HMM's many years ago, these techniques are just now accepted by the vision community. An early effort by Yamato et al. [29] uses discrete HMM's to recognize image sequences of six different tennis strokes among three subjects. This experiment is significant because it uses a 25x25 pixel quantized subsampled camera image as a feature vector. Even with such low-level information, the model can learn the set of motions and recognize them with respectable accuracy. Darrell and Pentland [5] use dynamic time warping, a technique similar to HMM's, to match the interpolated responses of several learned image templates. Schlenzig et al. [21] use hidden Markov models to recognize "hello," "good-bye," and "rotate." While Baum-Welch re-estimation was not implemented, this study shows the continuous gesture recognition capabilities of HMM's by recognizing gesture sequences. Closer to the task of this paper, Wilson and Bobick [28] explore incorporating multiple representations in HMM frameworks, and Campbell et. al. [2] use a HMM-based gesture system to recognize 18 T'ai Chi gestures with 98% accuracy.

3   Tracking Hands in Video

Previous systems have shown that, given some constraints, relatively detailed models of the hands can be recovered from video images [6,20]. However, many of these constraints conflict with recognizing ASL in a natural context, either by requiring simple, unchanging backgrounds (unlike clothing); not allowing occlusion; requiring carefully labelled gloves; or being difficult to run in real time.

In this project we have tried two methods of hand tracking: one, using solidly-colored cloth gloves (thus simplifing the color segmentation problem), and two, tracking the hands directly without aid of gloves or markings. Figure 1 shows the cap camera mount, and Figure 2 shows the view from the camera's perspective in the no-gloves case.

In both cases color NTSC composite video is captured and analyzed at 320 by 243 pixel resolution on a Silicon Graphics 200MHz Indy workstation at 10 frames per second. When simulating the self-contained wearable computer under development, a wireless transmission system is used to send real-time video to the SGI for processing [15].

In the first method, the subject wears distinctly colored cloth gloves on each hand (a pink glove for the right hand and a blue glove for the left). To find each hand initially, the algorithm scans the image until it finds a pixel of the appropriate color. Given this pixel as a seed, the region is grown by checking the eight nearest neighbors for the appropriate color. Each pixel checked is considered part of the hand. This, in effect, performs a simple morphological dilation upon the resultant image that helps to prevent edge and lighting aberrations. The centroid is calculated as a by-product of the growing step and is stored as the seed for the next frame. Given the resultant bitmap and centroid, second moment analysis is performed as described in the following section.

Fig. 1. The baseball cap mounted recognition camera.

In the second method, the the hands were tracked based on skin tone. We have found that all human hands have approximately the same hue and saturation, and vary primarily in their brightness. Using this information we can build an *a priori* model of skin color and use this model to track the hands much as was done in the gloved case. Since the hands have the same skin tone, "left" and "right" are simply assigned to whichever hand is currently leftmost and rightmost. Processing proceeds normally except for simple rules to handle hand and nose ambiguity described in the next section.

4   Feature Extraction and Hand Ambiguity

Psychophysical studies of human sign readers have shown that detailed information about hand shape is not necessary for humans to interpret sign language [18,22]. Consequently, we began by considering only very simple hand shape features, and evolved a more complete feature set as testing progressed [25].

Since finger spelling is not allowed and there are few ambiguities in the test vocabulary based on individual finger motion, a relatively coarse tracking system may be used. Based on previous work, it was assumed that a system could be designed to separate the hands from the rest of the scene. Traditional vision algorithms could then be applied to the binarized result. Aside from the position of the hands, some concept of the shape of the hand and the angle of the hand relative to horizontal seemed necessary. Thus, an eight element feature vector consisting of each hand's x and y position, angle of axis of least inertia, and eccentricity of bounding ellipse was chosen. The eccentricity of the bounding ellipse was found by determining the ratio of the square roots of the eigenvalues that correspond to the matrix

$$\begin{pmatrix} a & b/2 \\ b/2 & c \end{pmatrix}$$

Fig. 2. View from the tracking camera.

where a, b, and c are defined as

$$a = \int\int_{I'} (x')^2 dx' dy'$$

$$b = \int\int_{I'} x'y' dx' dy'$$

$$c = \int\int_{I'} (y')^2 dx' dy'$$

(x' and y' are the x and y coordinates normalized to the centroid) The axis of least inertia is then determined by the major axis of the bounding ellipse, which corresponds to the primary eigenvector of the matrix [8]. Note that this leaves a 180 degree ambiguity in the angle of the ellipses. To address this problem, the angles were only allowed to range from -90 to +90 degrees.

When tracking skin tones, the above analysis helps to model situations of hand ambiguity implicitly. When a hand occludes either the other hand or the nose, color tracking alone can not resolve the ambiguity. Since the nose remains in the same area of the frame, its position can be determined and discounted. However, the hands move rapidly and occlude each other often. When occlusion occurs, the hands appear to the above system as a single blob of larger than normal mass with significantly different moments than either of the two hands in the previous frame. In this implementation, each of the two hands is assigned the moment and position information of the single blob whenever occlusion occurs. While not as informative as tracking each hand separately, this method still retains a surprising amount of discriminating information. The occlusion event is implicitly modeled, and the combined position and moment information are retained. This method, combined with the time context provided by hidden Markov models, is sufficient to distinguish between many different signs where hand occlusion occurs.

5   Training an HMM Network

Unfortunately, space does not permit a treatment of the solutions to the fundamental problems of HMM use: evaluation, estimation, and decoding. A substantial body of literature exists on HMM technology [1,9,19,30], and tutorials on their use can be found in [9,24]. Instead, this section will describe the issues for this application.

The initial topology for an HMM can be determined by estimating how many different states are involved in specifying a sign. Fine tuning this topology can be performed empirically. While better results might be obtained by tailoring different topologies for each sign, a four state HMM with one skip transition was determined to be sufficient for this task (Figure 3). As an intuition, the skip

Fig. 3. The four state HMM used for recognition.

state allows the model to emulate a 3 or 4 state HMM depending on the training data for the particular sign. However, in cases of variations in performance of a sign, both the skip state and the progressive path may be trained.

When using HMM's to recognize strings of data such as continuous speech, cursive handwriting, or ASL sentences, several methods can be used to bring context to bear in training and recognition. A simple context modeling method is embedded training. Initial training of the models might rely on manual segmentation. In this case, manual segmentation was avoided by evenly dividing the evidence among the models. Viterbi alignment then refines this approximation by automaticaly comparing signs in the training data to each other and readjusting boundaries until a mimimum variance is reached. Embedded training goes on step further and trains the models *in situ* allowing model boundaries to shift through a probabilistic entry into the initial states of each model [30]. Again, the process is automated. In this manner, a more realistic model can be made of the onset and offset of a particular sign in a natural context.

Generally, a sign can be affected by both the sign in front of it and the sign behind it. For phonemes in speech, this is called "co-articulation." While this can confuse systems trying to recognize isolated signs, the context information can be used to aid recognition. For example, if two signs are often seen together, recognizing the two signs as one group may be beneficial. Such groupings of 2 or 3 units together for recognition has been shown to halve error rates in speech and handwriting recognition [25].

A final use of context is on the inter-word (when recognizing single character signs) or phrase level (when recognizing word signs). Statistical grammars relating the probability of the co-occurrence of two or more words can be used to weight the recognition process. In handwriting, where the units are letters, words, and sentences, a statistical grammar can quarter error rates [25]. In the absence of enough data to form a statistical grammar, rule-based grammars can effectively reduce error rates.

6 Experimentation

Since we could not exactly recreate the signing conditions between the first and second experiments, direct comparison of the gloved and no-glove experiments is impossible. However, a sense of the increase in error due to removal of the gloves can be obtained since the same vocabulary and sentences were used in both experiments.

6.1 Experiment 1: Gloved-Hand Tracking

The glove-based handtracking system described earlier worked well. In general, a 10 frame/sec rate was maintained within a tolerance of a few milliseconds. However, frames were deleted where tracking of one or both hands was lost. Thus, a constant data rate was not guaranteed. This hand tracking process produced an 16 element feature vector (each hand's x and y position, delta change in x and y, area, angle of axis of least inertia - or first eigenvector, length of this eigenvector, and eccentricity of bounding ellipse) that was used for subsequent modeling and recognition. Initial estimates for the means and variances of the

Table 2. Word accuracy of glove-based system

experiment	training set	independent test set
grammar	99.4% (99.4%)	97.6% (98%)
no grammar	96.7% (98%) (D=2, S=39, I=42, N=2500)	94.6% (97%) (D=1, S=14, I=12, N=500)

output probabilities were provided by iteratively using Viterbi alignment on the training data (after initially dividing the evidence equally among the words in the sentence) and then recomputing the means and variances by pooling the vectors in each segment. Entropic's Hidden Markov Model ToolKit (HTK) is used as a basis for this step and all other HMM modeling and training tasks. The results from the initial alignment program are fed into a Baum-Welch re-estimator, whose estimates are, in turn, refined in embedded training which ignores any initial segmentation. For recognition, HTK's Viterbi recognizer is used both with and without a strong grammar based on the known form of the sentences. Contexts are not used, since a similar effect could be achieved with the strong grammar given this data set. Recognition occurs five times faster than real time.

Word recognition accuracy results are shown in Table 2; the percentage of words correctly recognized is shown in parentheses next to the accuracy rates. When testing on training, all 500 sentences were used for both the test and train sets. For the fair test, the sentences were divided into a set of 400 training sentences and a set of 100 independent test sentences. The 100 test sentences were not used for any portion of the training. Given the strong grammar (pronoun, verb, noun, adjective, pronoun), insertion and deletion errors were not possible since the number and class of words allowed is known. Thus, all errors are vocabulary substitutions when the grammar is used (accuracy is equivalent to percent correct). However, without the grammar, the recognizer is allowed to match the observation vectors with any number of the 40 vocabulary words in any order. Thus, deletion (D), insertion (I), and substitution (S) errors are possible. The absolute number of errors of each type are listed in Table 2. The accuracy measure is calculated by subtracting the number of insertion errors from the number of correct labels and dividing by the total number of signs. Note that, since all errors are accounted against the accuracy rate, it is possible to get large negative accuracies (and corresponding error rates of over 100%). Most insertion errors correspond to signs with repetitive motion.

6.2   Analysis

The 2.4% error rate of the independent test set shows that the HMM topologies are sound and that the models generalize well. The 5.4% error rate (based on accuracy) of the "no grammar" experiment better indicates where problems may occur when extending the system. Without the grammar, signs with repetitive or long gestures were often inserted twice for each actual occurrence. In fact, insertions caused almost as many errors as substitutions. Thus, the sign "shoes" might be recognized as "shoes shoes," which is a viable hypothesis without a language model. However, a practical solution to this problem is the use of context training and a statistical grammar instead of the rule-based grammar.

Using context modeling as described above may significantly improve recognition accuracy in a more general implementation. While a rule-based grammar explicitly constrains the word order, statistical context modeling would have a similar effect while generalizing to allow different sentence structures. In the speech community, such modeling occurs at the "triphone" level, where groups of three phonemes are recognized as one unit. The equivalent in ASL would be to recognize "trisines" (groups of three signs) corresponding to three words, or three letters in the case of finger spelling. Unfortunately, such context models require significant additional training.

In speech recognition, statistics are gathered on word co-occurence to create "bigram" and "trigram" grammars which can be used to weight the likelihood of a word. In ASL, this might be applied on the phrase level. For example, the random sentence construction used in the experiments allowed "they like pill yellow they," which would probably not occur in natural, everyday conversation.

As such, context modeling would tend to suppress this sentence in recognition, perhaps preferring "they like food yellow they," except when the evidence is particularly strong for the previous hypothesis.

Unlike our previous study [23] with desk mounted camera, there was little confusion between the signs "pack," "car," and "gray." These signs have very similar motions and are generally distinguished by finger position. The cap-mounted camera seems to have reduced the ambiguity of these signs.

6.3   Experiment 2: Natural Skin Tracking

The natural hand color tracking method also maintained a 10 frame per second rate at 320x240 pixel resolution on a 200MHz SGI Indy. The word accuracy results are summarized in Table 3; the percentage of words correctly recognized is shown in parentheses next to the accuracy rates.

Table 3. Word accuracy of natural skin system

experiment	training set	independent test set
grammar	99.3% (99%)	97.8% (98%)
no grammar	93.1% (99%) (D=5, S=30, I=138, N=2500)	91.2% (98%) (D=1, S=8, I=35, N=500)

6.4   Analysis

A higher error rate was expected for the gloveless system, and indeed, this was the case for less constrained "no grammar" runs. However, the error rates for the strong grammar cases are almost identical. This result was unexpected since, in previous experiments with desktop mounted camera systems [23], gloveless experiments had significantly lower accuracies. The reason for this difference may be in the amount of ambiguity caused by the user's face in the previous experiments whereas, with the cap mounted system, the nose provided little problems.

The high accuracy rates and types of errors (repeated words) indicate that more complex versions of the experiment can now be addressed. From previous experience, context modeling or statistical grammars could significantly reduce the remaining error in the gloveless no grammar case.

7   Discussion and Conclusion

We have shown an unencumbered, vision-based method of recognizing American Sign Language (ASL). Through use of hidden Markov models, low error rates were achieved on both the training set and an independent test set without invoking complex models of the hands.

However, the cap camera mount is probably inappropriate for natural sign. Facial gestures and head motions are common in conversational sign and would cause confounding motion to the hand tracking. Instead a necklace may provide a better mount for determining motion relative to the body. Another possiblity is to place reference points on the body in view of the cap camera. By watching the motion of these reference points, compensation for head motion might be performed on the hand tracking data and the head motion itself might be used as another feature.

Another challenge is porting the recognition software to the self-contained wearable computer platform. The Adjeco ANDI-FG PC/104 digitizer board with 56001 DSP was chosen to perform hand tracking as a parallel process to the main CPU. The tracking information is then to be passed to a Jump 133Mhz 586 CPU module running HTK in Linux. While this CPU appears to be fast enough to perform recognition in real time, it might not be fast enough to synthesize spoken English in parallel (BT's "Laureate" will be used for synthesizing speech). If this proves to be a problem, newly developed 166Mhz Pentium PC/104 boards will replace the current CPU module in the system. The size of the current prototype computer is 5.5" x 5.5" x 2.75" and is carried with its 2 "D" sized lithium batteries in a shoulder satchel. In order to further reduce the obtrusiveness of the system, the project is switching to cameras with a cross-sectional area of 7mm. These cameras are almost unnoticeable when integrated into the cap. The control unit for the camera is the size of a small purse but fits easily in the shoulder satchel.

With a larger training set and context modeling, lower error rates are expected and generalization to a freer, user independent ASL recognition system should be attainable. To progress toward this goal, the following improvements seem most important:

- Measure hand position relative to a fixed point on the body.
- Add finger and palm tracking information. This may be as simple as counting how many fingers are visible along the contour of the hand and whether the palm is facing up or down.
- Collect appropriate domain or task-oriented data and perform context modeling both on the trisine level as well as the grammar/phrase level.
- Integrate explicit head tracking and facial gestures into the feature set.
- Collect experimental databases of native sign using the apparatus.
- Estimate 3D information based on the motion and aspect of the hands relative to the body.

These improvements do not address the user independence issue. Just as in speech, making a system which can understand different subjects with their own variations of the language involves collecting data from many subjects. Until such a system is tried, it is hard to estimate the number of subjects and the amount of data that would comprise a suitable training database. Independent recognition often places new requirements on the feature set as well. While the modifications mentioned above may be initially sufficient, the development process is highly empirical.

So far, finger spelling has been ignored. However, incorporating finger spelling into the recognition system is a very interesting problem. Of course, changing the feature vector to address finger information is vital to the problem, but adjusting the context modeling is also of importance. With finger spelling, a closer parallel can be made to speech recognition. Trisine context occurs at the sub-word level while grammar modeling occurs at the word level. However, this is at odds with context across word signs. Can trisine context be used across finger spelling and signing? Is it beneficial to switch to a separate mode for finger spelling recognition? Can natural language techniques be applied, and if so, can they also be used to address the spatial positioning issues in ASL? The answers to these questions may be key to creating an unconstrained sign language recognition system.

Acknowledgements

The authors would like to thank Tavenner Hall for her help editing and proofing early copies of this document.

References

1. L. Baum. "An inequality and associated maximization technique in statistical estimation of probabilistic functions of Markov processes." Inequalities, 3:1–8, 1972.
2. L. Campbell, D. Becker, A. Azarbayejani, A. Bobick, and A. Pentland "Invariant features for 3-D gesture recognition," Intl. Conf. on Face and Gesture Recogn., pp. 157-162, 1996
3. C. Charayaphan and A. Marble. "Image processing system for interpreting motion in American Sign Language." Journal of Biomedical Engineering, 14:419–425, 1992.
4. Y. Cui and J. Weng. "Learning-based hand sign recognition." Intl. Work. Auto. Face Gest. Recog. (IWAFGR) '95 Proceedings, p. 201–206, 1995
5. T. Darrell and A. Pentland. "Space-time gestures." CVPR, p. 335–340, 1993.
6. B. Dorner. "Hand shape identification and tracking for sign language interpretation." IJCAI Workshop on Looking at People, 1993.
7. I. Essa, T. Darrell, and A. Pentland. "Tracking facial motion." IEEE Workshop on Nonrigid and articulated Motion, Austin TX, Nov. 94.
8. B. Horn. Robot Vision. MIT Press, NY, 1986.
9. X. Huang, Y. Ariki, and M. Jack. Hidden Markov Models for Speech Recognition. Edinburgh Univ. Press, Edinburgh, 1990.
10. T. Humphries, C. Padden, and T. O'Rourke. A Basic Course in American Sign Language. T. J. Publ., Inc., Silver Spring, MD, 1980.
11. B. Juang. "Maximum likelihood estimation for mixture multivariate observations of Markov chains." AT&T Tech. J., 64:1235–1249, 1985.
12. W. Kadous. "Recognition of Australian Sign Language using instrumented gloves." Bachelor's thesis, University of New South Wales, October 1995.

13. C. Lee and Y. Xu, "Online, interactive learning of gestures for human/robot interfaces." IEEE Int. Conf. on Robotics and Automation, pp 2982-2987, 1996.

14. R. Liang and M. Ouhyoung, "A real-time continuous gesture interface for Taiwanese Sign Language." Submitted to UIST, 1997.

15. S. Mann "Mediated reality". "MIT Media Lab, Perceptual Computing Group TR# 260"

16. L. Messing, R. Erenshteyn, R. Foulds, S. Galuska, and G. Stern. "American Sign Language computer recognition: Its Present and its Promise" Conf. the Intl. Society for Augmentative and Alternative Communication, 1994, pp. 289-291.

17. K. Murakami and H. Taguchi. "Gesture recognition using recurrent neural networks." CHI '91 Conference Proceedings, p. 237–241, 1991.

18. H. Poizner, U. Bellugi, and V. Lutes-Driscoll. "Perception of American Sign Language in dynamic point-light displays." J. Exp. Pyschol.: Human Perform., 7:430–440, 1981.

19. L. Rabiner and B. Juang. "An introduction to hidden Markov models." IEEE ASSP Magazine, p. 4–16, Jan. 1996.

20. J. Rehg and T. Kanade. "DigitEyes: vision-based human hand tracking." School of Computer Science Technical Report CMU-CS-93-220, Carnegie Mellon Univ., Dec. 1993.

21. J. Schlenzig, E. Hunter, and R. Jain. "Recursive identification of gesture inputers using hidden Markov models." *Proc. Second Ann. Conf. on Appl. of Comp. Vision*, p. 187–194, 1994.

22. G. Sperling, M. Landy, Y. Cohen, and M. Pavel. "Intelligible encoding of ASL image sequences at extremely low information rates." Comp. Vision, Graphics, and Image Proc., 31:335–391, 1985.

23. T. Starner and A. Pentland. "Real-Time Ameerican Sign Language Recognition from Video Using Hidden Markov Models." MIT Media Laboratory, Perceptual Computing Group TR#375, Presented at ISCV'95.

24. T. Starner. "Visual Recognition of American Sign Language Using Hidden Markov Models." Master's thesis, MIT Media Laboratory, Feb. 1995.

25. T. Starner, J. Makhoul, R. Schwartz, and G. Chou. "On-line cursive handwriting recognition using speech recognition methods." ICASSP, V–125, 1994.

26. T. Takahashi and F. Kishino. "Hand gesture coding based on experiments using a hand gesture interface device." SIGCHI Bul., 23(2):67–73, 1991.

27. S. Tamura and S. Kawasaki. "Recognition of sign language motion images." Pattern Recognition, 21:343–353, 1988.

28. A. Wilson and A. Bobick. "Learning visual behavior for gesture analysis." Proc. IEEE Int'l. Symp. on Comp. Vis., Nov. 1995.

29. J. Yamato, J. Ohya, and K. Ishii. "Recognizing human action in time-sequential images using hidden Markov models." Proc. 1992 ICCV, p. 379–385. IEEE Press, 1992.

30. S. Young. HTK: Hidden Markov Model Toolkit V1.5. Cambridge Univ. Eng. Dept. Speech Group and Entropic Research Lab. Inc., Washington DC, Dec. 1993.

Towards Automatic Translation from Japanese into Japanese Sign Language*

Masaaki Tokuda and Manabu Okumura

School of Information Science,
Japan Advanced Institute of Science and Technology
(Tatsunokuchi, Ishikawa 923-12 Japan)
{tokuda,oku}@jaist.ac.jp

Abstract. In this paper, we present a prototype translation system named SYUWAN which translates Japanese into Japanese sign language. One of the most important problems in this task is that there are very few entries in a sign language dictionary compared with a Japanese one. To solve this problem, when the original input word does not exist in a sign language dictionary SYUWAN applies several techniques to find a similar word from a Japanese dictionary and substitutes this word for the original word. As the result, SYUWAN can translate up to 82% of words which are morphologically analyzed.

1 Introduction

The deaf communicate with each other with sign language composed of hand, arm and facial expressions. Japanese deaf use Japanese sign language (JSL) which is different from both phonetic languages (such as Japanese) and other sign languages (such as American Sign Language). According to recent linguistic research, JSL has a peculiar syntax. Unfortunately, there is little research on JSL and there are no practical machine translation (MT) systems that translate between JSL and Japanese.

Adachi et al.[1] analyzed the translation of parallel daily news corpora between Japanese and sign language, and studied a method to translate from Japanese into sign language. As a result they constructed useful translation rules. They supposed that all of the words in the daily news are recorded in the sign language dictionary. Nishikawa and Terauchi [2] studied the expression and translation of sign language based on computer graphics technologies. They designed and implemented a translation system, but they assumed that input words have already been analyzed and marked with all the information necessary for translation (part of speech, word sense, and so on). Fujishige and Kurokawa [3] studied a translation system that translates Japanese sentences into semantic networks. They showed that semantic networks are easy to use for generating

---

* This work is partly supported by the Inamura Foundation.

V. O. Mittal et al. (Eds.): Assistive Technology and AI, LNAI 1458, pp. 97–108, 1998.
© Springer-Verlag Berlin Heidelberg 1998

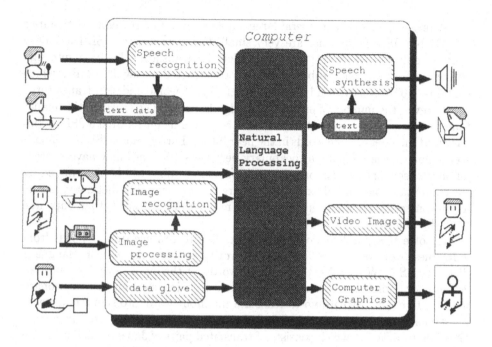

Fig. 1. An outline of MT system for sign language

a sign sentence. However, they also suppose that the sign language dictionary includes complete information about input words.

Sign language is very difficult to process on a computer because it is a visual language. Recently, there has been much research on recognition and generation of sign language [5] [6], but practical input/output devices or computerized systems of sign language are not available, and the processing load of sign language is very heavy. Moreover, they primarily studied computer graphics and used a complex description for sign language which can be easily processed by computers but is difficult for humans to understand. Therefore, a description for sign language which is easy for humans to understand and suitable for automatic processing is needed.

In this paper, we introduce "Sign Language Description Method (SLDM)" which is a description method of sign language using Japanese words as labels for expressing a word sequence. We propose a translation method similar to Adachi's work and implement a translation system named "SYUWAN" which processes raw input data of Japanese sentences and outputs the corresponding symbols based on SLDM.

We adopt a transfer approach as a translation method for SYUWAN which needs both source and target language dictionaries. In this work, we use a Japanese dictionary as the source language dictionary and a sign language dictionary as the target language dictionary.

At present, we can use several large Japanese machine readable dictionaries (MRD) but JSL dictionaries are very small. For instance, the JSL Dictionary contains only 634 words [11]. It is insufficient for translating Japanese into sign language. To solve this problem, we have constructed a new JSL MRD which contains 3162 words with advice from several deaf people and sign translators.

However, the number of entries in the sign language dictionary is still smaller than that of Japanese word dictionaries. It will be difficult to build a JSL dictionary that includes enough words in the near future. Furthermore, JSL vocabulary seems smaller than Japanese vocabulary because a JSL word may have a larger number of senses than a Japanese word, but its meaning will be determined by the context and nonverbal expression. Therefore, even if a JSL dictionary is perfectly constructed it would have a smaller number of head words than Japanese word dictionaries.

To solve this problem we propose a method to retrieve similar words from a Japanese word dictionary. If the input word does not exist in a sign language dictionary, SYUWAN substitutes similar words for the input word. This process is useful for finding an alternative word which exists in a sign language dictionary in order to enable the system to translate the input word into sign language.

We have also experimented with daily news sentences. These are taken from "NHK Syuwa News" which consists of translated pairs of Japanese and JSL. We analyze these Japanese sentences and translate them into SLDM expressions. If these words are not entries in a JSL dictionary, SYUWAN tries to get similar words and translates them to the JSL words.

In section 2, a description of sign language is given. The sign language dictionary of the MT system is described in section 3. The translation method of SYUWAN is illustrated in section 4. Section 5 describes some experimental results and problems of SYUWAN. Finally, conclusions and further work are discussed in section 6 and 7, respectively.

2   Description of Sign Language

Many description methods of sign language are proposed, but they are not suitable to be used in a computer because they mix pictures or special characters together [7] [8].

In this paper, we propose a simple but expressive description method named "Sign Language Description Method (SLDM)" which uses Japanese words and a few symbolic characters as labels. In this description, Japanese words and sign language words are related in a one-to-one manner [9] [10]. For example, "*Kyou, Hon Wo Katta* (I bought a book today.)" is denoted by "*Kyou / Hon / Kau* (today / book / buy)". Although we have to maintain link information between Japanese labels and Japanese sign language (JSL) expressions, the label of JSL words can be associated with original JSL expressions. The symbolic characters express direction, time and finger spelling (the Japanese syllabary). For example, the character "<" expresses pointing to something or hand direction. A SLDM sentence, "*Kare <*" (he <) expresses the action that the left hand takes the shape of "*Kare*" (he), and the right hand points to the left hand. The other SLDM sentence, "*Kare < Kiku*" (he < ask) expresses that the left hand takes the shape of "*Kare*" (he), and the right hand takes the shape of "*Kiku*" (ask) and moves to the left hand.

We translate Japanese into sign language based on SLDM. In this way, there is no need for computer intensive tasks such as image processing to display sign language using computer graphics. Moreover it is easy to map between sign language and SLDM expressions. In the future we will connect SYUWAN to other input/output devices.

Our system is based on a transfer method which translates an input sentence into a target language using predefined rules. This seems to work well because our sign language SLDM is very similar to Japanese language. In transfer methods, dictionaries of source and target languages, as well as a set of structure transfer rules, are necessary. In this work, the Japanese EDR[1] [15] [16] dictionary and a JSL dictionary are used as the source language dictionary and the target language dictionary, respectively. The EDR dictionary is a machine readable dictionary (MRD) which includes enough words for the Japanese language, but the JSL word dictionaries are very small, such as the one [11] that contains only 634 words. Although there are other dictionaries which contain up to 3000-6000 words [14] [13] [12], they include many useless words (dialect, obsolete words, and so on). To solve this problem, we collected Japanese head words from several JSL dictionaries, re-edited them and constructed a larger sign language word dictionary. The newly edited dictionary includes 3162 words. These words are common entries in several dictionaries and used in real situations. We think these words are sufficient for daily conversation.

3   Translation from Japanese into Sign Language

We implement a prototype Machine Translation system named "SYUWAN". It translates Japanese into JSL in a 4-step process as follows:

1. Japanese morphological analysis and removal of needless words.
2. Application of translation rules and direct translation to sign language words.
3. Translation to sign language words using similar words.
4. Translation to finger spelling.

Currently, our translation method fails to deal with the problem of word sense ambiguity; this problem remains as future work.

3.1   Morphological Analysis

In Japanese words in text are not bound by space. Therefore, it is necessary to determine a word boundary, part of speech and inflection. We use JUMAN [17] for morphological analysis which successfully processes 98% words. After

---

[1] Japan Electronic Dictionary Research Institute, Ltd.

this analysis, SYUWAN removes needless words (auxiliary verbs, etc.) from the output of JUMAN. We uses news sentences in the experiment of SYUWAN in this work because the structures of these sentences in Japanese and JSL look similar, where an auxiliary verb does not influence a structure of a sentence.

3.2   Direct Translation into Sign Language Words

In the next step, SYUWAN checks whether each word exists in the JSL word dictionary. If it does, the system translates it to the corresponding JSL word. SYUWAN applies the phrase translation rule for date expressions, and termination translation rules for the end of word (inflection) at the same time. The former rule is for a special expression in JSL, where the left hand expresses month and the right hand expresses the day at the same time. For example, if there is a date expression like "*9 gatsu 18 nichi*"[2] (September 18), SYUWAN applies a phrase rule and outputs "*/ 9 gatsu 18 nichi /*"[3]. The latter rule applies to the ends of the word which is translated into symbol characters. Because this kind of translation is difficult, we believe processing with rules is more suitable.

3.3   Finding a Sign Language Word Using Japanese Dictionary

One of the most important problems in translation between Japanese and JSL is the small size of Japanese label vocabulary in the JSL word dictionary. In the described experiment in the next section, half of the words in the daily news sentences are not included in the JSL dictionary. These words cannot be translated in the direct translation step. Therefore, SYUWAN tries to find JSL words from the Japanese word dictionary.

   If the input word does not exist in the JSL word dictionary, SYUWAN tries to get similar words from the machine readable Japanese word dictionary and translates them to JSL words. To derive similar words, three following methods are applied.

Using the Concept Identifier SYUWAN tries to get similar words which have the same concept identifiers with the input word from EDR Japanese Word Dictionary [15]. This dictionary includes nearly 400,000 word entries. Each entry is composed of the head word, the concept identifier, the definition sentence of a concept (Japanese and English), and so on. The head words which have the same concept identifiers are similar in the sense. Therefore, SYUWAN tries to get some head words which have the same concept identifiers as an original word, and tries to translate them into sign language word.

   For example, when SYUWAN translates the word "*gakushoku* (the refectory")" which is not an entry in the JSL dictionary it gets a concept identifier of "*gakushoku*" and head words which have the same concept identifiers as

---

[2] This is a Japanese expression.
[3] This is an SLDM expression.

"*gakushoku*". The results are "*shokudou* (a lunch counter)", "*byuffe* (a buffet)", "*ryouri-ya* (a restaurant)" and so on. The word "*shokudou*" is only an entry in the JSL word dictionary. Finally, SYUWAN translates the word "*gakushoku*" to "*shokudou*" (Figure 2).

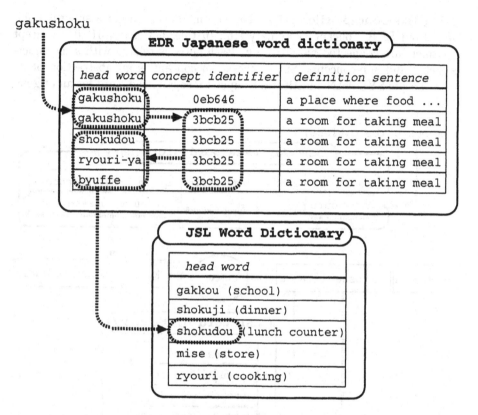

Fig. 2. Using the concept identifier

Using the Definition Sentence of a Concept: In the second method, the definition sentence of a concept is used. The EDR Japanese word dictionary has the definition sentence of a concept and a head word. SYUWAN extracts a list of words from a definition sentence using some extraction rules. These words are processed later in the same manner as the input. There are two kinds of extraction rules called "overlap-remove" rules and "*toiu*" rules. The former rule removes an overlap head word from the definition sentence. If the definition sentence contains the head word, SYUWAN processes that word recursively without termination. The overlap-remove rule avoids these situation. The latter rules are constructed from the expression of the EDR Japanese word dictionary. The definition sentence of a peculiar noun is usually in a fixed form like a "*Nippon toiu*

kuni" (A country called Japan). The "*toiu*" rule translates this explanation to
"*kuni / namae / Nippon*" (country / name / Japan).

   After the application of these rules, the derived explanation is processed by
morphological analysis, and a word list is extracted (Figure 3).

Using the Concept Hierarchy: The last method is to get a super-concept
word from EDR Concept Dictionary [16]. In this dictionary, 510,000 concept
identifiers are connected hierarchically. Each entry is related with some super-
concepts and sub-concepts. Since each super-concept is an abstraction of its
sub-concepts SYUWAN substitutes a super-concept's head word for an original

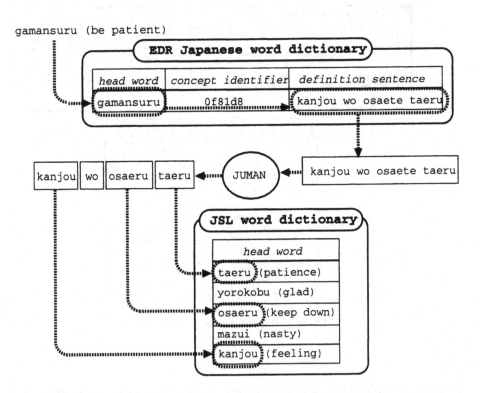

Fig. 3. Using concept explanation

word. For example, a "fish" concept connects "salmon", "sardine", "mackerel"
as its sub-concepts, and "organism" as its super-concept (Figure 4).

3.4   Translation into Finger Spelling

As the last step, SYUWAN translates the remaining words to finger spelling
if they are nouns. The finger spelling expresses Japanese syllabaries (Kana).

Each character of Japanese syllabaries corresponds to a unique finger shape. This method is used frequently in translation from Japanese nouns which do not exist in the JSL word dictionary, such as technical terms. Since the finger spelling has one-to-one correspondence with Japanese syllabary, the translation into JSL always succeeds. However, this method does not consider word senses and applies only for nouns at the last step of our system.

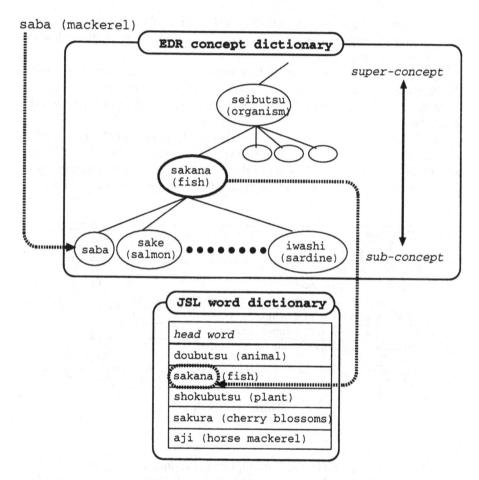

Fig. 4. Using hierarchical concept

4 Experiments and Results

In the experiment of SYUWAN we used daily news sentences as the input data. The news sentences are taken from NHK[4] news for people who are hearing

---
[4] Nippon Housou Kyoukai (Japan Broadcasting Corporation)

impaired. These news sentences consist of Japanese speech and sign language which are suitable for our work due to the following characteristics:

- It is easy to perform translation between Japanese and sign language because the newscaster (not a deaf person) translates the news sentence, and there are many similar constructions between both structures.
- We can get unusual sign language words (politics and economy terms, etc.).

We prepared 90 sentences including 1272 words. We corrected the results of morphological analysis and attach a certain concept identifier to each words beforehand.

SYUWAN processes an input in four steps as mentioned above. The result of the experiment is shown below. The "success" in the table means the translation succeeds and "failure" means SYUWAN processes the words, but the translation is incorrect.

We summarize the results of our experiments: 546 words (42%) were needless for SLDM at the first step of morphological analysis, 252 words (20%) were successfully translated at the step of direct translation to sign language word, and 59 words (5%) were successfully translated by translation rules. (Table 1)

Processing	success	failure	other
Needless words			546
Direct Translation	252	0	
Apply Translation rules	59	0	
Translate by the use of same concept identifier	64	4	
Translate by the use of definition sentence	91	137	
Translate by the use of super-concept	2	26	
Translate into finger spelling	37	35	
System failure		19	
Total	505	221	546

Table 1. Result

In the next step, SYUWAN attempts to get similar words from the EDR dictionary. At first SYUWAN got words which had the same concept identifiers, 68 words were found, and 64 (5%) words were translated successfully. Secondly SYUWAN got word lists from a definition sentence, and 228 words were found. After applying the extraction rules, there were 91 (7%) words which could be translated successfully. Lastly SYUWAN got super concepts of head words, as the result, and 28 words were found. We confirmed the result by hand, 2 (0.1%) words were translated successfully. In this step, 157 (=64+91+2) words could be successfully translated.

In the last step, among the remaining words, 37 words (3%) were translated to finger spelling, and 19 words (1%) could not be translated into sign language.

In conclusion, SYUWAN succeeded with up to 82% of the input words. Among 726 words, 252 words (34%) were directly translated, 59 words (8%) were translated by rule, and 157 words (22%) were translated by finding similar words. Therefore, SYUWAN could successfully translate up to 70% of the input words.

The translation failures occurred mainly in step 3. The following are some main reasons:

Useless definition sentences. In the EDR Japanese word dictionary, definition sentences of some concepts are useless. For example, the definition of *"saigen"* (reappearance) is *"saigen-suru"* (reappear), and *"mujitsu"* (not guilty). Its definition sentence is a *"tsumi ga mujitsu na koto"* (It is not guilty). Even if the extraction rules are applied, no useful words can be obtained from these sentences, then it is impossible to find similar words by this method. Since the EDR dictionary is going to be revised in the near future, we plan to run experiments with the new version of EDR dictionary and expect to get a better results.

Long definition sentence. Definition sentences of some concepts do not include appropriate words, and sometimes they are too long for SLDM. For the head word *"tagaku* (a lot of money)"*, the definition sentence is *"suuryou ya teido ga takusan de aru koto.* (a large quantity or a large number)"*. We prefer to get two words "money" and "many", but they are not included in the definition sentence.

It is possible to solve this problem by translating word explanations recursively, but it is not practical because noise would be added when SYUWAN processes a definition sentence of a concept. It is necessary to improve the processing precision of SYUWAN.

No entry in dictionaries. Some words like *"gokigenyou"* (How are you?) and *"dewa"* (as for) are in both the EDR dictionary or the JSL word dictionary. We think that this problem can be solved by adding more words in the JSL dictionary.

Kana expression. If the input sentence is notated by kana SYUWAN cannot recognize words in the sentence correctly. To solve the problem, we plan to use reading information in the EDR dictionary and try to describe the input with kanji.

5   Conclusions

We proposed a new description method for sign language named sign language Description Method which uses Japanese words as labels. We constructed a large JSL dictionary including 3162 head words, and implemented a MT system SYUWAN which translates Japanese into JSL. One of the most important problem in the MT of sign language is that there are very few entries in a sign language dictionary. For the case where a word does not exist in the sign language dictionary of an MT system we proposed the following techniques to find a target word based on similarity from the EDR Japanese word dictionary and translate into sign language:

- Translate into a sign language word which has the same concept identifier.
- Translate into sign language words using the definition sentence of a concept.
- Translate by the use of super-concept.

As the result of the experiment, our system can succeed up to 70 % of translation.

6 Future Work

We did not deal with word sense ambiguity in this work. It is not always the case that there is a one-to-one correspondence between a JSL word sense and a Japanese word sense. Methods to automatically determine word sense will be taken into account as our future work.

In this experiment, SYUWAN can translate 70% of the input words. This result is not satisfactory; we plan to improve SYUWAN to completely translate input to sign language.

In this paper, we proposed a translation method of word-to-word correspondence which tends to do a literal translation like Manually Coded Japanese (MCJ). SYUWAN is a prototype which can be used for a learning purpose or beginners. In future, we will consider the grammar of JSL in more and implement a more practical version of SYUWAN.

Acknowledgments

We are grateful to members of Okumura laboratory, especially Dr. Thanaruk Theeramunkong, for many comments and advice.

In construction of our dictionary, we received much help and many ideas from the deaf and sign translators in Ishikawa. Much help was obtained from a sign language circle in Nonoichi named TENOHIRA.

References

1. Hisahiro Adachi, Shouzou Yosizawa, Takashi Matsumoto, Masako Fujita, Kazuo Kamata : Analysis of News Sentences with Sign Language and Sign Translation Processing. Information Processing Society of Japan, Natural Language Processing. Vol.91-3 (1992) 17–24 (in Japanese).
2. Makoto Nishikawa, Mutsuhiro Terauchi : A Study on Japanese - Sign Language Translation Considering Placement of Sign Words. IEICE, Human Communication Science, Vol.94-94,(1995) 55–62 (in Japanese).
3. Fujishige Eiichi, Kurokawa Takao : Semantic Network Generation from a Sign Word Sequence and its Translation to a Japanese Sentence. 10th Human Interface Tokyo. (1994) 103–108 October (in Japanese).

4. Takao Kurokawa : Gesture Coding and a Gesture Dictionary for a Nonverbal Interface. IEICE Trans. Fundamentals, Vol.E75-A, No.2. (1992) 112–121. February.

5. Jintae Lee, T.L.Kunii : Generation and Recognition of Sign Language Using Graphic Models. IISF/ACM Japan International Symposium: Computers as Our Better Partners. (1994) 96–103. March.

6. Eiji Ohira, Hirohiko Sagawa, Tomoko Sakiyama, Masaru Ohki: A Segmentation Method for Sign Language Recognition. IEICE Trans. Inf. & Syst. E78 D, (1995) 49–57, January.

7. Kanda Kazuyuki : Syuwagaku kougi. Fukumura Shuppan. (1994) (in Japanese).

8. Honna Nobuyuki, Kato Mihoko : Syuwa no Hyouki-hou ni tsuite. Sign Language Communication Studies of The Quarterly of The Japan Institute for Sign Language Studies, No.4. (1990) 2–9. (in Japanese).

9. McCoy K.F., McKnitt W.M., Peischl D.M., Pennington C.A., Vanderheyden P.B., Demasco P.W. : AAC-user therapist interactions : Preliminary linguistic observations and implications for companision. Proc. of the RESNA '94 Annual Conference. (1994) 129–131.

10. Japanese Federation of the Deaf : Sign Language for the Intermediate Class. (1990) (in Japanese).

11. Kanda Kazuyuki : Japanese Sign Language dictionary. Alpha Media. (1994).

12. Syuwa communication Kenkyukai : Shin Syuwa Jiten. Chuuouhouki.(1992).

13. Maruyama Kouji : Irasuto Syuwa Jiten. Dynamic Sellers Publishing Co.,Ltd. (1984).

14. Japanese Federation of the Deaf : The glossaries of the Japanese Signs (1987) (Watashi tachi no SYUWA in Japanese).

15. Japan Electronic Dictionary Research Institute LTD : Japanese Word Dictionary Version 1.0.(1995).

16. Japan Electronic Dictionary Research Institute LTD : Concept Dictionary Version 1.0. (1995).

17. Nagao Lab, Kyoto University : Japanese Morphological Analysis System JUMAN Manual version 2.0.(1994).

An Augmentative Communication Interface Based on Conversational Schemata

Peter B. Vanderheyden[1]* and Christopher A. Pennington[2]

[1] Department of Computer Science
University of Waterloo
Waterloo, Ontario N2L 3G1
Canada
pbvander@uwaterloo.ca
[2] Applied Science and Engineering Laboratories
Department of Computer and Information Sciences
University of Delaware / duPont Hospital for Children
Wilmington, DE 19899
U.S.A.
penningt@asel.udel.edu

Abstract. Many people with severe speech and motor impairments make use of augmentative and alternative communication (AAC) systems. These systems can employ a variety of techniques to organize stored words, phrases, and sentences, and to make them available to the user. It is argued in this chapter that an AAC system should make better use of the regularities in an individual's conversational experiences and the expectations that the individual normally brings into a conversational context.

An interface and methodology are described for organizing and retrieving sentences appropriate to a particular conversational context, sentences that were possibly developed from earlier conversations. These conversations are represented according to the schema structures discussed by Schank as a model for memory and cognitive organization [16]. The interface allows the user to proceed with minimal effort through conversations that follow the schema closely, and facilitates the derivation of new schemata when the conversation diverges from existing ones. This interface, called SchemaTalk, is intended to operate in parallel with and to complement a user's existing augmentative communication system.

The results of preliminary investigations into the effectiveness of the interface and methodology have been encouraging; further investigations are planned. Of interest for future study is how the use of schematized text might influence the way that augmented communicators are perceived by their conversational partners. Possible design modifications to improve the usability of the interface are also under investigation.

---

* Much of the work upon which this chapter is based was completed while the first author was at the University of Delaware Center for Applied Science and Engineering, located at the duPont Hospital for Children, in Wilmington, DE.

V. O. Mittal et al. (Eds.): Assistive Technology and AI, LNAI 1458, pp. 109–125, 1998.
© Springer-Verlag Berlin Heidelberg 1998

1   Introduction

The goal of designing a system for augmentative and alternative communication (AAC) is to facilitate the communication of people who have difficulty with speech, writing, and sign language. Speech synthesis and digitally-encoded recorded speech have made it possible to provide people with a new voice. Yet for the most part, people with severe speech and motor impairments operate their communication devices using adapted keyboards and other computer-oriented input devices — these can be difficult to use, and the resulting sentence production can be very slow. The design of the interface to the AAC system can go a long way towards enabling the user, the augmented communicator, to exercise this new voice more effectively and with less effort.

Early AAC devices were simply boards or books containing symbols (letters, words, and/or pictures). The person using the communication board pointed at a symbol on the board, and another person — typically the conversational partner, the person with whom the communication board user was conversing — was responsible for identifying the symbols and interpreting their meaning. The burden of interpretation was laid on this second person, as was the power to control the conversation and to manage the topic and turns. This loss of conversational control by the AAC user could lead to a perception of reduced communicative competence [13] and a lower sense of social empowerment.

Computerized AAC systems can do more than just provide the augmented communicator with a new voice. They can organize words and sentences to be more easily accessible. They can apply natural language techniques, making use of lexical, syntactic, and semantic information within the current sentence ([8], and Pennington (this volume)) to predict their user's next word or to fill in missing words. These are a few of the ways in which computerized AAC systems can support the novel production of well-formed sentences, requiring less time and effort from the augmented communicator.

However, there may often arise situations when it is not necessary for the augmented communicator to construct a novel sentence, conversations that proceed as one would anticipate on the basis of prior experience. In this chapter, we explore the representation of prior conversational text as yet another method for facilitating an augmented communicator's participation in day-to-day conversations. The augmented communicator is able to organize this text around the greater conversational context, thus building a convenient and natural platform for interpersonal interaction.

Every day, each of us takes part in any number of activities, and in any number of brief or extended conversations. Some of these are with familiar people and in familiar locations, others are with people we've never met before, and yet for the most part we carry off these interactions successfully. How do we do it? In each situation do we, somewhere in the back of our minds, generate a novel response to each stimulus? This seems unlikely; if we had to think about every action from scratch, we might never get anything done at all. Instead, we can often base our actions on something we've done before, or something we've said before.

This kind of reusability of experience is very practical and comes naturally to augmented and unaugmented communicators alike, with one significant difference: the amount of physical work required to reproduce an earlier utterance is considerably greater for an augmented communicator than for an unaugmented one. When unaugmented communicators recall from an earlier experience something to say, they say it (at rates that may be in the range of 200 words/minute); when augmented communicators recall from an earlier experience something to say, they may need to reconstruct it from scratch on their AAC system (at rates that may be in the range of 2-10 words/minute).

Clearly then, one step to increasing the rate of augmented communication could be for the AAC system to support the retrieval of ready-to-use text from previous conversations. But retrieval is not the complete answer. To facilitate competent communication, the retrieved text must be appropriate to the current situation, and retrieval should be easier and faster than reconstructing the text from scratch. Efficient retrieval of appropriate text will be encouraged, we believe, by organizing text in a manner that is both consistent and intuitive to the augmented communicator.

In the next section, we describe the motivation behind our approach to organizing conversational text as scripts and schemata. This is followed by a brief review of relevant work in AAC. Then we present our SchemaTalk interface, describing the preliminary implementation and investigative studies. Finally, possible extensions to SchemaTalk are discussed, as are several ongoing initiatives.

2 Representing the Structure and Content of Conversation

The idea that we can model the structure and content of many everyday experiences is not new, and we will describe one general approach for doing so. This approach has been applied to understanding stories and news articles, to performing common activities such as preparing a meal, as well as to representing conversation.

2.1 Scripts of Familiar Events

When we are told that a customer has picked up some clothes from the neighborhood laundromat, we immediately make certain assumptions: the clothes belong to the customer, the clothes were in fact cleaned, the customer paid for the cleaning, etc. Why is it that we infer these details if they were not explicitly stated? Schank and Abelson's [15] answer to this question is that we are familiar with the activity of picking up clothes at the laundromat as a result of having done it many times, and we have each developed a mental script to represent the sequence of events involved — typically, for example, the clothes were clean when we came to pick them up and we paid for them. Many scripts are built up during the course of an individual's lifetime, as a result of one's experiences and the interpretations of those experiences.

Processing information by incorporating it into scripts has a number of advantages. The sheer magnitude of the information is reduced by storing only once every occurrence that follows the typical sequence of events. Only when an exception occurs (for example, we notice the spot on a favorite shirt has not been removed) do we store the details of that event. By representing the typical sequence of events for a given situation, scripts also provide a means of inferring actions that have not yet taken place or have not been explicitly stated. A script for a familiar situation can also be abstracted and used to provide initial expectations in a related but novel situation. To give an example, if we take our clothes to an entirely different laundromat that we have never visited before, we can generalize our experiences with the familiar laundromat to apply to this new one.

2.2 Schemata for Stories

Later, Schank ([16]; informally in [17]) extended and modified this idea of linear scripts into a hierarchy of schema structures, including memory organization packets (MOPs) and metaMOPs. Let us continue with the laundromat example and take the abstraction a level or two higher. In any novel situation where we are the customer, we would expect to pay for services rendered. A metaMOP would capture this general situation in Schank's system — metaMOPs represent high-order goals. The metaMOP for the pay-for-services goal would contain separate MOPs for more specific instances of this goal, such as picking up clothes from the laundromat, or picking up the car from the mechanic. MOPs can themselves be hierarchical, so a general laundromat MOP can be associated with any number of MOPs for specific laundromats. Each MOP contains scenes, or groups of actions, that occur within that MOP. The MOP for picking up clothes at the laundromat might include an entrance scene, a scene for getting the clothes, a scene for paying, and so on. Each scene has associated with it any number of scripts, where a script contains the actual actions that have taken place. One script for the entrance scene, for example, may include opening the door, walking in, and greeting the shopkeeper.

To demonstrate the use of schemata in understanding stories and answering questions, Schank described a number of computer experiments including the CYRUS system [16]. CYRUS contained databases of information about two former Secretaries of State, integrated new information with these databases, and provided answers to questions such as "Have you been to Europe recently?" and "Why did you go there?" More recently, the DISCERN program developed by Miikkulainen also answered questions based on input texts [14]. It represented schemata subsymbolically, in terms of features and probabilities rather than words.

2.3 Schemata for Conversations

The schemata in the question-answering systems of Schank and Miikkulainen would represent the locutionary information that is exchanged during a conversation, i.e., its literal content. A model of conversation should also consider the intention and form of utterances, as well as the overall structure in which the exchange of utterances occurs between the participants. Schemata can be used also for this purpose.

Kellermann et al. [11] described conversations between students meeting for the first time as a MOP and identified 24 scenes. These conversations appeared to have three phases: initiation, maintenance, and termination. Scenes in the initiation phase, for example, involved the participants exchanging greetings, introducing themselves, and discussing their current surroundings. Scenes tended to be weakly ordered within each phase but strongly ordered between phases, so that a person rarely entered a scene in an earlier phase from a scene in a later phase. A number of scenes involved what the investigators called subroutines, or the common sequence of generalized acts: get facts, discuss facts, evaluate facts, and so on.

JUDIS [18] was a natural language interface for Julia [7], an interactive system that played the part of a caterer's assistant and helped the user plan a meal. JUDIS operated on goals, with each goal represented as a MOP containing the characters (caterer and customer), scenes (either mandatory or optional), and the sequence of events. Higher-level MOPs handled higher-level goals, such as the goal of getting information, while lower-level MOPs handled lower-level goals, such as answering yes-no questions. Recognizing that the person with whom it was interacting had goals of their own, JUDIS tried to model those goals on the basis of the person's utterances. It also recognized that several MOPs could contain the same scene, and several scenes could contain the same utterance. Only one MOP was executed at a time, but other MOPs consistent with the current state in the conversation would be activated.

3   Augmentative Communication Systems

Augmentative communication systems must address the abilities and needs of their users, and we review strengths and weaknesses of a number of traditional alternatives, below. AAC systems should also consider the contexts in which the systems will be used, and this includes the context of conversation.

3.1   Letter-, Word-, and Sentence-Based Systems

Some AAC systems are letter- or spelling-based, requiring the user to enter words letter by letter. Letter-based systems can have many of the strengths and weaknesses of word-based systems. Letter-based input is flexible, potentially removing even the constraints imposed by system vocabulary limits, since any sequence of letters can potentially be entered by the user. However, the demands of entering each letter can be even greater than the demands of entering whole words. This is one reason why some letter-based systems attempt to predict the word as it is being entered, reducing some demands on the user but possibly introducing others [12].

With a word-based interface, the user selects individual words and word-endings. Such systems also offer the advantage of a great deal of flexibility. The user can produce any sentence for which the vocabulary is available, and is in complete control of the sentence content, length, and form. However, effective word-based sentence production relies heavily on manual dexterity or access rate, and on the individual's linguistic and cognitive abilities. An individual who can select only one item per minute will either produce very short sentences or will produce long gaps in a conversation while selecting the words. Such a system may not be suited to an individual who has difficulty generating well-formed or appropriate sentences [9].

Sentence-based systems allow an individual to utter an entire sentence by selecting a single key sequence, resulting in much faster sentence production. The sentence that is produced can be prepared to be long or short, and linguistically well-formed, thus overcoming some difficulties of word-based systems. However, strict sentence-based systems have shortcomings of their own. The user is limited to the often small number of sentences prestored in the system. These sentences may be syntactically correct, but are cumbersome to modify to be appropriate to a given semantic or pragmatic context. As well, the user can incur additional cognitive load if the interface design makes the task of locating and retrieving sentences non-trivial [4].

Of course, systems need not be exclusively letter-based, word-based, or sentence-based. On a system from the Prentke Romich Corporation called the Liberator^{TM}, for example, a user can map an icon key sequence to a word, a phrase, or an entire sentence. Templates can also be set up containing a phrase or sentence with gaps to be filled by the user at the time of utterance.

3.2 Conversational Considerations

Several features of conversation differentiate it from prepared speech or writing. Silent pauses are strongly discouraged, while discontinuities and grammatical errors are often overlooked; the topic of conversation can change suddenly, or be interrupted and then continued. How could AAC systems be designed to support this kind of dynamic participation? Two existing systems, CHAT and TOPIC, offer interesting suggestions.

CHAT [2] was a prototype communication system that recognized general conversational structure. A model conversation would begin with greetings, move on to smalltalk and the body of the conversation, then finally to wrap-up remarks and farewells. Often, the exact words we use for a greeting or a farewell are not as important as the fact that we say something. A person using CHAT could select the mood and the name of the person with whom they were about to speak, and have CHAT automatically generate an appropriate utterance for that stage of the conversation. An utterance was chosen randomly from a list of alternatives for each stage. Similarly, while pausing in our speech to think or while listening to the other participant in a conversation, it is customary to occasionally fill these gaps with some word or phrase. CHAT could select and utter such fillers quickly on demand.

To assist the augmented communicator during the less predictable main body of the conversation, a database management system and interface called TOPIC [3] was developed. The user's utterances were recorded by the system and identified by their speech acts, subject keywords, and frequencies of use. If the user selected a topic from the database, the system suggested possible utterances by an algorithm that considered the current semantics in the conversation, the subject keywords associated with entries in the database, and the frequency with which entries were accessed. The possibility of allowing the user to follow scripts was also considered.

These systems offered the user an interface into a database of possible utterances, drawn either from fixed lists (CHAT) specific to several different points in the conversation, or reusing sentences from previous conversations (TOPIC) organized by semantic links. Once the conversation had entered the main body phase, however, there was no representation of the temporal organization of utterances. As well, topics were linked in a relatively arbitrary network, rather than organized hierarchically.

The body of a conversation is not always as difficult to predict as these systems may imply by their design. There are many contexts in which conversations can proceed more or less according to expectations, as the designers of CHAT and TOPIC are now exploring in their script-based ALADIN project [1].

3.3 Communication Skills Training

One area where the structured and predictable conversations that can be represented in simple scripts have proven useful is in modeling context-appropriate communication behavior. Elder and Goossens' [9] proposed script-based strategies for training developmentally delayed adolescents and adults to use their augmentative communication systems in the contexts of domestic living, vocational training opportunities, leisure/recreation, and community living. They developed an activity-based communication training curriculum, in which students were taught context-appropriate communication in the process of performing the relevant activity with the instructor or with another student. A script was generated for each activity, and was represented by an overlay to be placed over the individual's AAC system. Supplemental symbols could be added off to the side of an overlay; for example, specific food types in a "making dinner" script. Activities that had a logical sequence were advantageous because one event in the activity could act as a cue to recall the next event, and it was impossible to successfully complete the activity in any but the correct order.

4   A Schema-Based AAC Interface

CHAT organized prestored utterances according to their place in the structure of a conversation, and TOPIC organized them by their semantic relations — an AAC system that represents prestored conversations as schemata can do both. Sentences are arranged into linear scripts, with the user advancing through the list as the conversation proceeds; scripts are combined to form scenes, and scenes are combined to form MOPs, always maintaining the expected structure of the conversation. At the same time, the hierarchical representation of the schemata reflects their semantic relations: only scenes that would occur within the context of a MOP are included in it, and only MOPs semantically related to one another are linked together. SchemaTalk [19] represents conversations in just this way.

The goal of SchemaTalk, as of any AAC device, is to assist an augmented communicator to participate as easily and effectively in conversation as possible. SchemaTalk has been implemented as an interface to an AAC system rather than as a system in itself to keep to a minimum the training that an augmented communicator would need before using it for the first time. It can either be accessed directly using a computer mouse or keyboard, or it can be accessed through its user's regular AAC system.

When a conversation proceeds in the expected manner, the SchemaTalk interface provides schemata containing relevant prestored sentences for the current portion of the conversation. This is done by allowing the user to select the MOP (or topic, see below) of the conversation at which point SchemaTalk will show a list of scenes each containing sentences that can be selected. SchemaTalk displays these sentences and allows the user to proceed through the schemata. In addition to complete sentences, a scene in SchemaTalk may contain sentence and phrase templates which allow the user to fill in small amounts of conversation-specific information, producing complete sentences in a short amount of time. These templates provide additional flexibility with little effort from the user. Thus, the interface is intended to provide the speed of access inherent in many sentence-based systems, but with greater flexibility.

4.1   The Schema Framework

The different schema and template structures available on the SchemaTalk interface might best be described using a number of examples. For the most part, we will consider the context of going to a restaurant for dinner.

A separate MOP is planned out for each conversational context that is likely to reoccur frequently, and for which there are reasonably well-developed expectations. These MOPs are then grouped by similar goals, and a higher-level MOP is developed by generalizing among the members of each group. For example, MOPs for going to McDonald's, Burger King, and the lunchroom cafeteria might all be grouped under the "eating at a self-serve restaurant" MOP. Going to other restaurants might fall under the "eating at a waiter-served restaurant" MOP, and together these two MOPs would be contained in the more general "eating at a restaurant" MOP. Subordinate structures such as scenes and scripts are associated with the appropriate MOPs.

The preceding paragraph suggests a top-down approach to designing schemata; the remainder of this subsection will consider a bottom-up approach. When an augmented communicator enters a fine dining establishment such as the Blue and Gold Club at the University of Delaware, they would expect to be greeted by a maître d'. The exchange might go something like this:

Maître d': "Good evening. Under what name is the reservation?"
Customer: "Vanderheyden."
Maître d': "Ah yes, Mr. Vanderheyden. Will that be a table for..."
Customer: "A table for two, please, by the window."
Maître d': "Very good sir. Please follow me."

If the customer were composing a MOP for this experience, this exchange could well be considered part of a restaurant entry scene and the customer's two utterances could be stored as the script for this scene.

Having been seated, and after a polite amount of time during which the customer perused the menu, the maître d' might return:

Maître d': "Could I bring you a drink before the meal?"
Customer: "A glass of white wine, please."
      or "A glass of water, please."
      or "Tea, please."
    ⋮

During this second scene, the customer may be accustomed to giving one of several replies to the maître d's question. Any number of these replies could be represented verbatim for this scene in SchemaTalk. A second alternative exists, however. Rather than representing many sentences that share a similar form and occur in the same place in a conversation, the sentence form containing the common elements could be stored only once, and the elements that vary could be linked to it. Thus, the template "_____, please." could be stored, with the blank slot fillable by "A glass of wine," or "A glass of water," etc.

Thus an ordered sequence of scenes, each scene containing a list of scripts (currently no more than one), and each script containing an ordered list of sentences or sentence templates, can be stored as a MOP for eating dinner at a restaurant.

When the augmented communicator tries to use this MOP in a very different restaurant situation, perhaps a visit to the neighborhood McDonald's, the scenes and sentences may not apply very well. The previously developed MOP might be relabeled as the MOP for eating at the Blue and Gold, or more generally for eating at waiter-served restaurants, and a new MOP for eating at McDonald's would need to be developed. Either or both of these MOPs could be used to develop a more general MOP for eating at a generic restaurant.

This McDonald's MOP might contain an entry scene with no sentences at all, the first sentences occurring only in the scene for ordering the meal. In this order scene, the sentences "I'd like a shake and an order of fries" and "I'd like a Big Mac and a root beer" could be produced with the template "I'd like a _____" and the fillers "shake" and "order of fries" or "Big Mac" and "root beer ."[1] Using the template, only three selections are required to produce either sentence, rather than nine selections (or six, if "order of fries" is counted as one word) to enter the words on a word-based system. The templates also capture the intuitive similarities in the form and function of the sentences.

The hierarchical structure of this representation quite naturally leads to the idea of inheritance of properties by lower-level schemata. If "eating at McDonald's" involves the same sequence of scenes as contained in the more general "eating at a fast food restaurant," for example, it would access these scenes by inheritance from the more general MOP rather than containing a redundant and identical copy of them. The McDonald's MOP could still differ from the more general MOP by having its own set of fillers for slots (such as the list of food items that can be ordered) and its own set of scripts and sentences.

Inheritance provides a mechanism to provide schemata for contexts that are novel but similar in some respects to existing schemata. When the user enters a new fast food restaurant for the first time, if a MOP for a restaurant serving similar food exists, then it could be selected. If such a MOP does not already exist, then the MOP for the general fast food restaurant could be selected. Entering a restaurant for the first time, the interface is able to provide the user with an appropriately organized set of sentences and sentence templates to use.

4.2 Implementation Details

The SchemaTalk interface closely follows the framework described above. In describing how an augmented communicator would actually use the interface, we once again take the context of eating dinner at a restaurant as our example (Figure 1). Please note that, for readability, the term "MOP" has been replaced in SchemaTalk by the term "topic."

In order to bring up the appropriate set of scenes for the current conversation, the user navigates through a hierarchically-arranged set of topics. As a scene begins, the first sentence in its script is highlighted. The user may select this sentence by pressing the button labeled "Select" (when using the interface directly from a computer) or by pressing "Tab" key (when interfacing from the user's own AAC system). The sentence can be modified by the user and, when ready, spoken by pressing the button labeled "Speak" or by pressing the "Return" key. The interface uses the AAC system's own speech production facility to utter the sentence.

Selecting a sentence causes the next sentence to be highlighted, so that the user may proceed through a conversation scene that follows the script with minimal effort. The user may also, at any time, decide to produce sentences out of their scripted order by using the "Next" and "Prev" (previous) sentence buttons or their associated keys.

When the last sentence in a scene is spoken, the next scene is begun — the interface continues to keep pace with the conversation. Should the user decide

---

[1] The conjunction "and" is inserted automatically when multiple fillers are selected for a single slot.

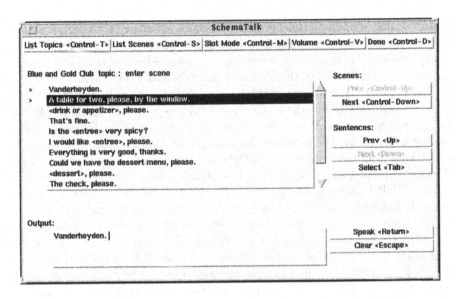

Fig. 1. The SchemaTalk interface, displaying sentences contained in the "Blue and Gold Club" restaurant topic

to break even further from the prepared schema, the SchemaTalk interface allows the user to change the current scene or even to change the current topic. The "List Scenes" and "List Topics" buttons call up dialog boxes (Figure 2) to perform precisely this operation. The user scrolls through the list of scenes or topics and makes a selection. The next and previous scenes can also be selected directly from the main window using the "Next" and "Prev" scene buttons or their associated keys.

In addition to the "Scene Mode" in which sentences were displayed in scene order (Figure 1), SchemaTalk also supports a more interactive and flexible alternative for constructing sentences from templates. This "Slot Mode" is entered automatically when a sentence template is selected, or when the "Slot Mode" button is pressed. Here the user can select words or phrases from a list that can be used to fill template slots.[2] Figure 3 shows the list of "entree" slot fillers that appears when the "Are the _____ very spicy?" template (which occurred as the fifth line in Figure 1) is selected. Interface operations in this mode parallel those in the scene mode, with slots and fillers accessible in the main window, and slots and topics also accessible as dialog boxes. Slot fillers can be selected from prepared lists or can be inputted directly from the user's AAC system. When multiple slot fillers are selected, the verb in the template sentence is automatically inflected and the slot fillers are separated by commas and the word "and" as appropriate.

---

[2] In the list of sentences, a slot is indicated by angle brackets ("< ...>"). The name of the slot appears within the angle brackets.

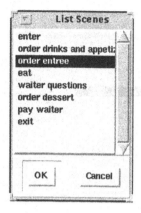

Fig. 2. Example "List Scenes" and "List Topics" dialog boxes

4.3   Preliminary Investigations

The goal of this interface is to facilitate an augmented communicator's participation in conversations. Preliminary investigations were held in order to analyze more closely the interface's effectiveness towards this goal and to get feedback on the SchemaTalk interface from people who use AAC systems regularly. In the experiment, two people used the SchemaTalk system in a mock job interview situation. During preliminary investigations, the two subjects played the role of job applicants and were recorded in separate conversations with four and three (respectively) other subjects playing the role of interviewers. Of the two applicant-subjects, one made use of an AAC system for his daily communication needs while the other was not an AAC user.

In an iterative fashion, the SchemaTalk developer guided each applicant-subject in developing a small number of simple preliminary schemata. The actual sentences and schemata that the subjects used in the SchemaTalk system were developed by the subject on the basis of the first several interviews. The applicant-subjects then employed these schemata in interviews and the interviews were recorded by the communication device. The developer and the applicant-subjects together reviewed these recorded interactions, and enhanced the existing schemata or developed new ones. A schema development program is planned for the future (please see the Future Work section, below), to allow the users to continue refining and adding schemata to the interface on their own.

Results were encouraging: both applicant-subjects produced utterances in less time (Figure 4) and at higher words per minute rates (Figure 5) when making use of schematic text than when entering novel text directly from their AAC systems (labeled on the graphs as "schema" and "direct," respectively).

This suggests that SchemaTalk is indeed effective at supporting a more efficient and active involvement by the augmented communicator in conversational situations.

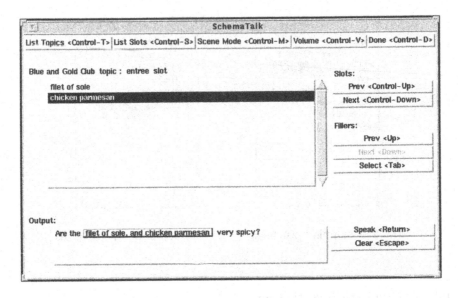

Fig. 3. The SchemaTalk interface in "Slot Mode"

Briefly, two other patterns were suggested by the investigation results. The number of turns taken by the applicant-subjects varied considerably across interviews, as did the number of words per utterance; whereas the former seemed strongly influenced by the characteristics of the interviewer, the latter seemed the result of strategies of the AAC user. The applicant-subject who used an AAC system regularly showed a preference to retrieving longer utterances from the schemata. The applicant-subject not accustomed to using an AAC system, however, preferred to retrieve shorter, more pragmatic utterances from schemata (perhaps demonstrating the conversational style towards which the CHAT system was aimed). It would be interesting to see whether such a dichotomy of patterns continues in further investigations.

The SchemaTalk interface is intended to be applicable to all AAC users and all conversational contexts. For this reason, it is hoped that a larger, more diverse group of people will eventually be able to participate in subsequent investigations. Of particular interest will be people who use their AAC systems in the context of their employment.

5   Future Work

The opportunity for reflection naturally brings with it suggestions for how SchemaTalk and the evaluation study could be improved.

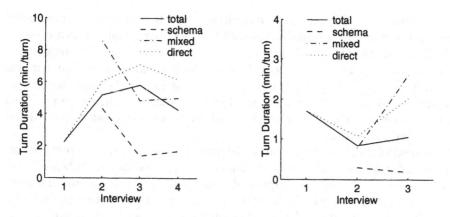

Fig. 4. Time per turn (minutes) for the two applicant-subjects

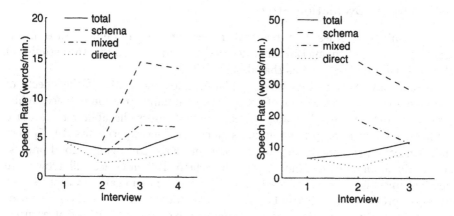

Fig. 5. Words per minute for the two applicant-subjects

5.1 Possible Extensions to the Interface

AAC systems that attempt to predict words on the basis of the initial letter(s) selected by the user in a domain-nonspecific context may have a very large vocabulary to consider for each word in a sentence. A similar problem of scale can face systems that attempt to complete partial or telegraphic sentences. A schema-based interface makes use of the current MOP and the current position within the MOP to define a specific conversational domain. This domain could serve to constrain or prioritize the vocabulary and semantics that the system would need to consider, and reduce the time to process the sentence.

The network of MOPs and their substructures must currently be constructed by the investigator, in consultation with the user. Determining which contexts should be represented is obviously a highly subjective issue that reflects the individuality of one's experiences. It would be preferable to develop a means by which users could construct their own hierarchy of schemata. Better still would be a dynamic system that could store sentences as they were produced during a conversation, and for which the schemata could be created and updated interactively by the user.

Several of these weaknesses in the interface presented in this chapter are currently being addressed. In a new version now under development, SchemaTalk II [6], the user can create new scripts and modify existing ones, and all commands can be executed with one or two keystrokes. The presentation of scripts in a simple "flat" list with MOP relations transparent to the user is being investigated. It is expected that these improvements will allow the augmented communicator to develop and access schematized text with greater ease.

5.2 Further Evaluation Studies

The preliminary studies described earlier were intended to demonstrate the effectiveness of the SchemaTalk interface design, and certainly further studies of this sort will be conducted as that design evolves.

In addition, SchemaTalk offers a platform for the study of the effects of reusable text on the perception of AAC users' communicative competence. While prestored text should allow augmented communicators the ability to produce complete context-appropriate sentences quickly, earlier experiments [10] [5] were inconclusive on the importance of message length for conveying a good impression of the augmented communicator. The SchemaTalk interface will allow more detailed investigation into the effects of specific communication variables. For example, prestored scripts can be varied in the amount of information they contain that is relevant to the conversational context, in the length of utterances, etc.

6 Summary

An interface for augmentative communication systems is described that makes the expected content of a conversation available to the user. This can facilitate interaction in predictable situations by reducing the need to produce common utterances from scratch. A methodology is given for organizing conversations in a variety of contexts according to hierarchical schema structures. Each MOP is related to a functional goal defined by the augmented communicator, and contains a list of scenes in the order in which they are expected to occur in the conversation. Each scene contains sentences that the augmented communicator can choose. Sentences can be complete or in the form of templates containing slots to be filled in as needed. Preliminary investigations reinforce the conclusion that this interface makes it possible to participate more easily in a conversation that fits a MOP closely, while also allowing the individual to enter novel sentences directly using their regular AAC system.

7 Acknowledgments

The authors would like to thank Dr. Kathleen McCoy for her very helpful comments, and Terry Carpenter for his clarification of current issues in the development of SchemaTalk II. This work has been supported by a Rehabilitation Engineering Research Center Grant from the National Institute on Disability and Rehabilitation Research of the U.S. Department of Education (#H133E30010). Additional support has been provided by the Nemours Foundation.

References

1. N. Alm, A. Morrison, and J.L. Arnott. A communication system based on scripts, plans, and goals for enabling non-speaking people to conduct telephone conversations. In *Proceedings of IEEE Systems, Man and Cybernetics*, pages 2408–2412, Vancouver, Canada, 1995.

2. N. Alm, A.F. Newell, and J.L. Arnott. A communication aid which models conversational patterns. In *Proceedings of the RESNA 10th Annual Conference*, pages 127–129, San Jose, CA, 1987.

3. N. Alm, A.F. Newell, and J.L. Arnott. Database design for storing and accessing personal conversational material. In *Proceedings of the RESNA 12th Annual Conference*, pages 147–148, New Orleans, LA, 1989.

4. B. Baker. Minspeak. *Byte*, pages 186–202, 1982.

5. J. L. Bedrosian, L. Hoag, D. Johnson, and S. N. Calculator. Communicative competence as perceived by adults with severe speech impairments associated with cerebral palsy. Manuscript submitted for publication, 1997.

6. T.L. Carpenter, Jr., K.F. McCoy, and C.A. Pennington. Schema-based organization of reusable text in AAC: User-interface considerations. In *Proceedings of the 20th Annual RESNA Conference*, pages 57–59, Pittsburgh, PA, 1997.

7. R. E. Cullingford and J. L. Kolodner. Interactive advice giving. In *Proceedings of the 1986 IEEE International Conference on Systems, Man and Cybernetics*, pages 709–714, Atlanta, GA, 1986.

8. P.W. Demasco and K.F. McCoy. Generating text from compressed input: An intelligent interface for people with severe motor impairments. *Communications of the ACM*, 35(5):68–78, 1992.

9. P. S. Elder and C. Goossens'. *Engineering Training Environments for Interactive Augmentative Communication: Strategies for adolescents and adults who are moderately/severely developmentally delayed*. Clinician Series. Southeast Augmentative Communication Conference Publications, Birmingham, AL, 1994.

10. L. Hoag, J. L. Bedrosian, D. Johnson, and B. Molineux. Variables affecting perceptions of social aspects of the communicative competence of an adult AAC user. *Augmentative and Alternative Communication*, 10:129–137, 1994.

11. K. Kellermann, S. Broetzmann, T.-S. Lim, and K. Kitao. The conversation MOP: Scenes in the stream of discourse. *Discourse Processes*, 12:27–61, 1989.

12. H. H. Koester and S. P. Levine. Quantitative indicators of cognitive load during use of a word prediction system. In *Proceedings of the RESNA '94 Annual Conference*, pages 118–120, Nashville, TN, 1994.
13. J. Light. Interaction involving individuals using augmentative and alternative communication systems: State of the art and future directions. *Augmentative and Alternative Communication*, 4(2):66–82, 1988.
14. R. Miikkulainen. *Subsymbolic Natural Language Processing: An integrated model of scripts, lexicon, and memory*. MIT Press, Cambridge, MA, 1993.
15. R. C. Schank and R. P. Abelson. *Scripts, plans, goals and understanding: An inquiry into human knowledge structures*. Erlbaum, Hillsdale, NJ, 1977.
16. R.C. Schank. *Dynamic Memory: A theory of reminding and learning in computers and people*. Cambridge University Press, NY, 1982.
17. R.C. Schank. *Tell Me A Story: A new look at real and artificial memory*. Charles Scribner's Sons, NY, 1990.
18. E.H. Turner and R.E. Cullingford. Using conversation MOPs in natural language interfaces. *Discourse Processes*, 12:63–90, 1989.
19. P. Vanderheyden. Organization of pre-stored text in alternative and augmentative communication systems: An interactive schema-based approach. Master's thesis, University of Delaware, Newark, DE, 1995.

Assistive Robotics: An Overview

David P. Miller

KISS Institute for Practical Robotics
10719 Midsummer Drive
Reston, VA 20191
U.S.A.
dmiller@kipr.org

Abstract. Intelligent robotics is the study of how to make machines that can exhibit many of the qualities of people. This seems a very appropriate technology to use to assist those people who have lost certain abilities that are common to the majority of the population. This paper gives an overview of some of the robotic technology that has been or is being developed to assist people with disabilities.

1 Introduction

All disabilities affect both the mind and body. While most human disabilities can be traced to injury or defect in the neural or information processing system, the affect on the body is usually in the form of a limitation on a person's sensory or motor capabilities.

Robotics is the blending of sensing, movement and information processing. This paper, and those that follow it will discuss how robotic systems can be used to supplement a person's own sensory and motor capabilities.

There are numerous places where a robotic system can be of assistance to a person. Some of those areas of application are:

- Assisting people who are mobility impaired:
  - Helping a person move from place to place
  - Bringing a person desired objects from a remote location
- Automated manipulation
  - Remote manipulation of objects
  - Allowing a person to feed themselves
- Guiding the sensory impaired
  - Translating sensory modalities
  - The automated "guide dog"

The remainder of this chapter will examine each of these robotic applications. We will survey some of the work that has been done in these areas. We will also examine the AI/robotic issues that are involved, and try and point out the tall poles for future work. The chapters that follow this one will provide detailed case studies for many of these technology areas.

V. O. Mittal et al. (Eds.): Assistive Technology and AI, LNAI 1458, pp. 126–136, 1998.
© Springer-Verlag Berlin Heidelberg 1998

2   Mobility

Mobility has been a mainstay of robotic research for decades. Aside from the mechanical and engineering issues of how to move a robot across the surface of the planet[1], mobility has proven a fruitful area for researching many issues of behavior and intelligence.

Unlike many applications of fixed robotic arms, where the environment is completely artificial and the initial conditions are completely known, mobile robots – almost by definition – leave the nest to explore parts unknown. Mobile robotic systems have had to deal with obstacle, navigation issues, and of course – figuring out where they want to go.

The latter has always proven rather sticky to calculate in any seemingly intelligent way. Fortunately, when assisting people, figuring out where to go is one of those issues that can best be left to the person the robot is attempting to assist.

For people who are mobility impaired there are two types of goals where assistance is particularly needed: getting the person to location X; and getting the person to within reach of item Y. The first of these almost always involves moving the person. The latter task can have more flexibility in its solution.

To move a mobility impaired person, a variety of mechanical solutions have been devised over the past five thousand years. Minor impairments can often be overcome through the use of canes and crutches while more major difficulties are usually solved by a manual wheelchair. A wheelchair is usually powered by the user or by a human assistant. However, people who wish to be independent, but are not capable of pushing a chair themselves have had the options of powered scooters or motorized wheelchairs for the last several decades.

For most people, the solutions above are adequate to get them to where they need to go, or to get them near the things they need to get near. Yet there are still hundreds of thousands of people who cannot safely manipulate any of the devices mentioned above. It is for these people that robotic assistance may prove very beneficial.

2.1   Going to the Mountain

Power wheelchairs are traditionally used by people who do not have the upper body strength and dexterity to operate a manual wheelchair. However, operation of a power wheelchair can still be a difficult and demanding task for many such individuals. The operator must be able to accurately sense their environment, recognize hazards, and be able to translate their mobility desires into continuous joystick commands for the chair.

A variety of user interfaces have been created to aid people that lack the ability to operate a traditional joystick in using power wheelchairs. In most instances, this involves repositioning the joystick and adding a mechanical attachment to the end of the joystick so that it may be operated by a person's elbow, chin,

---

[1] Or, as in a few special cases, across the surfaces of other planets [15].

or tongue. In some cases, an eye tracker is used, or options are flashed one at a time on a display, and the operator makes their   selection by pressing a single switch, blinking, or altering their breathing pattern. But in all these instances, the command options are basically the same: move forward or backwards, turn left or right.

For most operators who cannot use the traditional joystick, and even for many who can, operating the chair is a tedious, unnatural and stressful activity. Their limited bandwidth of interaction with the chair limits the speed at which they can safely travel, and often puts them in situations that are hazardous for themselves and for the objects in their environment. Additionally, many potential power wheelchair users have limited visual acuity, or must be seated in a way that limits their forward vision. None of the traditional interfaces address these vision related problems.

For many of these people, a smart wheelchair may be the solution to the their mobility needs. A smart wheelchair can sense its environment and the user's mobility desires and combine the two into a smooth and safe set of movements.

Wheelchairs as Robots In many ways, wheelchairs are ideal robots. Wheelchair manufacturers long ago solved the issues of reliable mechanics, motors, control electronics, etc. Wheelchairs can carry twice their weight in payload, travel several miles on a charge, operate every day, and cost (compared to most research robot platforms) a relatively modest amount.

What chairs lack as robots is a programmable interface and a method of integrating sensors. Neither of these problems is all that difficult to overcome, and several independent research projects have developed robotic wheelchairs with which to conduct their work.

Since 1993, KISS Institute has developed a series of robotic wheelchairs called the TinMan chairs which it sells to universities at cost in order to promote research in assistive robotics [18]. As a result of this program, the number of universities doing research on smart wheelchairs has more than doubled.

The TinMan wheelchair robots [16,17] use a commercial microcontroller and a set of interface electronics to interface with standard wheelchair controllers using the standard wheelchair controller's normal peripheral protocols. In other words, the supplementary controller on a TinMan chair appears to the chair controller as a joystick and to the joystick it appears as the chair controller. In this way, user commands can still be entered through the joystick or other interface device. The supplementary controller then processes the input and passes on new commands to the chair in the form of a string of simulated joystick movements. The supplementary controller also interfaces to any sensors that have been added to the robot. The sensor input is also used in generating the wheelchair commands.

The idea of mixing control between a users input through a joystick, and the navigation system of a reactive robot is well established in the literature. Connell [7] describes a robot called *Mr. Ed* which can be ridden by a person.

While the point of Connell's system was to make a robot that behaved more like a horse than a wheelchair, the concept is similar.

In the late '80s, a semi-robotic wheelchair [22] was used to help motivate children confined to a wheelchair. In [2,3,27], an ongoing research program that has produced a robotic wheelchair with capabilities similar to that of the Tin Man chairs is described. All of these systems, and others [10,11], are capable of simple obstacle avoidance.

In [17] we showed that smart wheelchairs can greatly reduce the amount of required operator interaction to guide the chair. By monitoring the joystick and other input devices during navigation tasks run in both manual and 'smart' mode we were able to evaluate the number of required operator interactions. It was also possible to evaluate the time criticality and accuracy of each each interaction. In all cases, the smart wheelchair needed much less input at a substantially lower degree of fidelity than an ordinary wheelchair. It should be noted that the operators for these tests were not mobility impaired, and that they were able to operate the chair at a slightly higher speed in manual mode than in its 'smart' mode.

Robots as Wheelchairs There are robots that you can sit on that conceivably can take you where you want to go. Now you just need to tell them where to go. There have been numerous approaches to this problem with no solution yet being completely satisfactory.

The most popular method is still to use the joystick, or a joystick compatible device as the interface between the user and the robotic wheelchair. Whether it is a joystick, a head tracker, an eye follower or some other device of this sort, the difficulty is in expressing a plan or a goal through a medium that was designed for describing real-time actions.

All of these devices have some neutral setting (the center position on a joystick) and then there is positive or negative movement along two axis. When a normal power chair operator wants to go down the hall and turn into the second open door on the left, then they push the joystick forward until they reach the second doorway; they then move the joystick to the left until they are lined up with the door. They then move the joystick forward again to enter the room.

Most robotic wheelchairs need the same steps from their operators as does a normal chair. The difference is that the robotic chairs will keep the chair centered in the hall, stop or go around an obstacle, and greatly ease the alignment process of getting through the doorway. So for someone with severe spasticity, or slow reactions, or limited visual acuity, or reduced peripheral vision these chairs can be a big win. Yet this is still a far cry from having someone in the chair somehow communicate to the chair that they want to go into the second room on the left.

Until recently, this was a difficult problem for mobile robots in general, but in the last few years there have been several public examples of robots doing this type of task using only the type of symbolic map that could be derived from

the statement "go down the hall and into the second doorway on the left" [12][2].
Now all the pieces exist and we can hope to see systems in the next few years
that can take voice (or some other modality with similar expressive capabilities)
commands of a plan/goal nature rather than strictly of a real-time action nature.

2.2   Having the Mountain Come to You

When you want to go to the Grand Canyon you really have no option other than
having your body somehow transported to the vicinity of the Grand Canyon.
The robots in the previous section may aid in that task. Alternatively, if you
want to see the movie "Grand Canyon" you can go to where it is being shown,
or you can have the video and video player moved to where you are. This latter
method is the realm of the *fetch-it robot*.

Fetch-it robots are both easier and more difficult to create then the smart
wheelchairs. They are easier because the task is more strictly defined and the
interface more straightforward. They are more difficult to create because more
of the task must be done autonomously. In particular, once the robot has been
tasked with retrieving a specific item (e.g., a particular video tape) then it is
up to the robot to decide where that item should be, planning and executing
movements to get it there, and then recognizing and acquiring the object once
the robot has reached the object's location.

Once again, non-AT-work in mobile robots has addressed all of these issues
and some recent systems have made great strides at integrating all of these capabilities. For example, the Kansas State entry at a recent mobile robot contest [25]
was able to find and retrieve several common household objects in an ordinary
living room setting.

Of course this and similar systems still have difficulty finding objects they
have never seen before that a human would not find difficult (e.g., finding the
remote for the CD player is easy for a person, because they know what a generic
remote looks like, and can, from reading the function labels on the buttons,
distinguish a remote directed at a CD player from one designed for a TV or
VCR). These robots also have trouble finding and retrieving objects that are
placed in particular awkward locations (e.g., the remote has fallen between the
cushions on the sofa).

However, for a well defined list of rigid objects that are located in reasonable
places, this is a solvable problem. If the robot is in charge of not only retrieving
the objects, but also of returning them to their storage locations, then these
robot systems can actually operate quite efficiently – since the search for the
objects can be greatly reduced.

---

[2] Though as of this writing, the author is unaware of any robot system that actually
creates its map on the fly from the description of the task.

3 Let Your Robot Do the Fingers

If a robot can find your TV remote and bring it back to you then there are a number of other manipulation tasks that robots ought to be able to do for people who have difficulty manipulating objects in the world. This section outlines some of the manipulation tasks on which there is active work.

3.1 Manipulation Assistance

If a person is mobility impaired, they may be able to move about the world in a wheelchair of one sort or another. However, wheelchairs put a person lower than a typical person when standing. A wheelchair also limits a person's ability to lean over or reach down. So most people in wheelchairs find it difficult to access objects that deviate much from table height. For some chair users, all that is needed is a simple mechanical aid – a grasping stick. But the people who are using the smart chairs discussed earlier probably require a more advanced aid.

There has been quite a bit of work on attaching robotic arms to wheelchairs. While the goal of the arm is clear: to pick and place objects out of the reaching capability of the chair user; how to achieve these goals is less clear.

A arm attached to a wheelchair must have at least six degrees of freedom. Any less will require that the chair be maneuvered while the arm is grasping, to assure an acceptable approach angle. Controlling those six degrees and the gripper and any speed and torque settings can be a clumsy and sometimes daunting task for any robot operator. The standard teach pendant, switchable joystick or spaceball interfaces are usually difficult to manage. If the operator is someone who is partially visually impaired, or whose view is occluded by the chair or their own body, or suffers spasticity, etc. – then the operation of a traditional arm can be quite impossible.

The control interface has been the major roadblock in the creation of a useful general purpose accessory arm for the wheelchair user. This problem has proven more daunting than the mass, power and cost issues – which themselves are critical. Fortunately, some progress has been made.

Pook [23,24] has designed a system that allows a user to point and issue commands such as "pick up that cup" or "open that door." The robot still has some difficulty figuring out how to grasp things, but can ask the user for a starting grasp strategy. Here the tradeoff is between how extensive the onboard sensor processing of the environment should be when compared to the level of autonomy of the arm activities. Activities such as "pull out that thing stuck under the chair" are still beyond the scope of this type of system because the object is unknown to the system and the user can offer no help. However, simple solutions such as adding a camera to the arm so that the user can see what needs to be done might solve even this class of problem in the near future.

3.2 Feeding Assistance

One of the areas of manipulation that most people really want to be able to handle themselves is eating (personal hygiene is another even more difficult and important area for automation). Fortunately, it is possible to use a somewhat structured environment to greatly simplify this problem compared to the general manipulation problem.

Food can can be put into a compartmentalized tray. Preprogrammed movements by a robot arm can then be used to scoop up a mouthful of food from a given compartment regardless of the position of the food in the compartment (e.g., [30]. If the tray is mounted to the user's chair, or to a table to which the chair is 'docked', then movements by a robot from the food tray to the vicinity of the user's mouth can also be preprogrammed.

The final approach must be a more closed-loop process. It must take into account both the position of the user's mouth and its state: whether the person is chewing or not; if ready to accept more food, whether the mouth is open or closed. An interface must also be included to give the user some control over what food is being scooped up and at what pace. The user needs to be able to open their mouth to speak without having a spoonful of applesauce automatically inserted. [13,1,6,28] describe systems that are currently being developed to accomplish the robotically assisted eating task.

4   "Watch out! Can't you hear that tree?"

Robot systems have always used slightly different sensor modalities than have ordinary humans. In most cases, the sensors and the associated processing are far inferior to their organic counterparts. Yet in some cases (such as distance ranging) robotic sensor systems are much more accurate than those available to unaugmented humans.

Artificial sensors have been used for years to help people. Reading machines [14] read printed text and output it in an audible form.[3] Submarine sonar systems convert audible echoes into a visual image as does a medical ultrasound scanner.

All of these system could be considered robotic or not. But the systems and capabilities discussed in the remainder of this section came from work in robotics. The prototypes for some of these systems were lifted directly from a mobile robot and strapped to a human with some sort of added interface.

4.1   Blind as a Bat?

The most common use of robot sensors is the use of ultrasonic sonar to aid people who are blind. Ultrasonic sonars have been used as a ranging device on robots for years. Several systems have been developed to help someone who is blind interpret the sonar to help them avoid obstacles and find their way through cluttered environments.

The *Mowat* [29] and *NOD* [4] sensors are two handheld sonars that translate echo time into vibratory or auditory frequencies respectively. The increasing

---

[3] Reading machines are, in most ways, robots. They sense and take physical action, are programmable, and react to the sensory input given them.

vibration or higher pitch warn users of a nearby obstacle in the direction that device is pointed.

Other devices such as the *Navbelt* [26] and *Sonicguide* [29] are worn and have a wide field of view, these devices report to the user both the distance and bearing to the various obstacles in the environment. By having stereo audio output, these devices are able to let the user hear the direction from which the echo is being reported.

All these devices require quite a bit of experience to accurately interpret and can interfere with normal hearing of environmental noises. These devices also have limited ability to pick up small obstacles at floor level, and therefore in many instances they act as a supplement to a cane.

4.2 Man's Best Friend

The most elegant solution for getting around in an urban environment for a person who is blind is the guide dog. These trained canines act not only as sensors but as processing system, expert advisors, high quality interface and companion. The dogs not only see a potential hazard, but evaluate it and come up with a strategy to go around it while still achieving their master's navigational goals. The person and dog interact through a capable interface that involves subtle physical pressures and a few audible commands and responses.

One obvious application of a robot system would be to replace a guide dog with a robotic analog. The benefits of such a replacement would be to reduce the maintenance required by the owner (while well-trained, a guide dog still needs to be fed, walked and have regular medical check-ups and maintenance). Dogs also are inconvenient in some locations and some people are allergic to them. More importantly, trained guide dogs are in short supply and very expensive to produce.

Two research projects that are creating robotic seeing eye dogs are the *Guide-Cane* [5] and *Hitomi* [20,21].

Hitomi is a system that can be used to automate most of the capabilities of a trained guide dog. This system combines satellite navigation, obstacle avoidance and clever vision techniques to help guide the user safely to their desired destination.

One of the most interesting features of this system is its use of vision to detect dangerous traffic situations. The robot tracks the shadows underneath a car to help isolate one car from another. At night, it tracks the lights of the vehicles.

Hitomi is built upon a power wheelchair and has the user take the position of someone who would be pushing the chair. The seat of the chair has been replaced by the electronic equipment used to guide the device. *Hitomi* is a very capable device using several different sensor modalities to ensure the user's safety. Unfortunately, it has all the access problems common with wheelchairs, and some additional problems because small bumps and soft ground cannot easily be detected until the device is right on top of them.

The *GuideCane* is a much smaller device, with some reduction in capability when compared to *Hitomi*. The *GuideCane* is really a small, sonar guided mobile robot with two drive wheels. The robot's balance is maintained by a handle which is held by the user. As an obstacle is approached, the robot turns to avoid it and giving the user a tactile warning through the twist in the handle. Like *Hitomi*, the *GuideCane* can use internal maps to plot a course and guide the user to it. Both devices have wheel encoders and can maneuver by dead reckoning when necessary. The *GuideCane* lacks vision or any absolute positioning system, however it is a much smaller device that can actually be lifted over bumps by the user. Unfortunately, doing so will eliminate the device's dead reckoning capability.

5   Summary

Robotic systems are potentially an important part of the assistive technology array of tools. But with few exceptions, there is little that can be done by a robot system in the area of assistive technology that could not be done more reliably by a human. The drawbacks of having a human assistant to guide a quadraplegic person, or someone who is blind, or to feed someone with no use of their arms, is the lack of independence and the cost. Most people prefer to be able to do common tasks (such as eating, moving around their home or office, or taking a walk through their neighborhood) by themselves. While most enjoy companionship, they tend to resent the dependence. In almost all cases, increasing the independence of a person is a very positive action.

The technical issues in getting these robotic systems to reach a useful level — one where the users could exist independently from outside human assistance — are the same problems that have dominated most of intelligent robotics research for the last decade: how to best integrate reactive and deliberative planning. Most of the systems described above are highly reactive — and have trouble interpreting the long range goals of their users. Some of the systems do integrate sequences of actions [8] in response to certain stimuli. However, currently very few robotic systems can fully use all the power available in a hybrid reactive/deliberative architecture [9]. While not reaching the full potential of intelligent assistive robotics, these systems are still able to be of great utility to many people who have disabilities that affect their sensory or motor capabilities.

But because of the structure of health care, at least in the United States, in order for this technology to be readily available to the people who need it, it must prove to be more cost effective than traditional solutions – or in fact no solution at all. Many people do not face the alternative of a robotic assistant or a human one, but rather the prospect of no personal assistance at all – life bedridden, or living in a full time care institution.

It is the goal of robotic assistive technology to free as many people as possible from life in an institution or as a shut-in. As this technology is made more effective and lower in cost it should be possible to allow thousands of people to take better care of themselves and make an active contribution to the rest of society.

References

1. S. Bagchi and K. Kawamura. ISAC: a robotic aid system for feeding the disabled. AAAI Spring Symposium on Physical Interaction and Manipulation, March, 1994.
2. D.A. Bell, S.P. Levine, Y. Koren, L.A. Jaros, and J. Borenstein. Shared control of the NavChair obstacle avoiding wheelchair. In *Proceedings of RESNA '93*, June 1993.
3. D.A. Bell, S.P. Levine, Y. Koren, L.A. Jaros, and J. Borenstein. Design criteria for obstacle avoidance in a shared-control system. In *Proceedings of RESNA '94*, June 1994.
4. D. Bissitt and A.D. Heyes. An application of biofeedback in the rehabilitation of the blind. *Applied Ergonomics*, vol 11 #1, pp 31-33.
5. J. Borenstein and I. Ulrich. The GuideCane — a computerized travel aid for the active guidance of blind pedestrians. Unpublished White Paper, University of Michigan, Advanced Technology Lab, 9/21/96.
6. M. E. Cambron and K. Fujiwara Feeding the physically challenged using cooperative robots. *Proceedings of the RESNA '94 Conference*, Nashville, TN, June, 1994.
7. J. Connell and P. Viola. Cooperative control of a semi-autonomous mobile robot. In *Proceedings of the 1990 IEEE Conference on Automation and Robotics*, pp. 118-1121, May 1990.
8. R. J. Firby. Adaptive execution in complex dynamic worlds. Ph.D. thesis, Yale University Department of Computer Science, January 1989.
9. E. Gat. Reliable goal-directed reactive control of autonomous mobile robots. Ph.D. thesis, Virginia Polytechnic Institute Department of Computer Science, April 1991.
10. T. Gomi and A. Griffith. Developing intelligent wheelchairs for the handicapped. In Mittal et al., [19]. This volume.
11. R. Gelin, J. M. Detriche, J. P. Lambert, and P. Malblanc. The sprint of coach. In *Proceedings of the '93 International Conference on Advanced Robotics*, November 1993.
12. D. Hinkle, D. Kortenkamp and D. Miller. The 1995 IJCAI robot competition and exhibition. *AI Magazine*, volume 17, #1 , 1996.
13. K. Kawamura, S. Bagchi, M. Iskarous and M. Bishay. Intelligent robotic systems in service of the disabled. *IEEE Transactions on Rehabilitation Engineering*, Vol. 3, pp. 14-21, March 1995.
14. K. Lauckner and M. Lintner. Computers: inside & out, fifth edition. Pippin Publishing, Ann Arbor, MI, 1996. Chapters 6-7.
15. D. P. Miller, R.S. Desai, E. Gat, R. Ivlev, and J. Loch. Reactive navigation through rough terrain: experimental results. In *Proceedings of the 1992 National Conference on Artificial Intelligence*, pp. 823-828, SanJose, CA, July 1992.
16. D. P. Miller and E. Grant. A robot wheelchair. In *Proceedings of the AAIA/NASA Conference on Intelligent Robots in Field, Factory, Service and Space*, March 1994.
17. D. P. Miller and M. Slack. Design & testing of a low-cost robotic wheelchair. *Autonomous Robots*, volume 1 #3, 1995.
18. D. P. Miller The TinMan wheelchair robots. http://www.kipr.org/robots/tm.html, KISS Institute for Practical Robotics.

19. V. Mittal, H.A. Yanco, J. Aronis and R. Simpson, eds. *Lecture Notes in Artificial Intelligence: Assistive Technology and Artificial Intelligence.* Springer-Verlag, 1998. This volume.

20. H. Mori, S. Kotani and N. Kiyohiro. A robotic travel aid "HITOMI." In Proceedings of the Intelligent Robots and Systems Conference, IROS 94, pp. 1716-1723, Munich, Germany, September, 1994.

21. H. Mori, S. Kotani and N. Kiyohiro. HITOMI: Design and development of a robotic travel aid. In Mittal et al., [19]. This volume.

22. P.D. Nisbet, J.P. Odor, and I.R. Loudon. The CALL Centre smart wheelchair. In *Proceedings of the First International Workshop on Advanced Robotics for Medical Health Care*, Ottawa, Canada, June 1988.

23. P. K. Pook. Teleassistance: using deictic gestures to control robot action. TR 594 and Ph.D. Thesis, Computer Science Dept., U. Rochester, September 1995.

24. P. Pook. Saliency in human-computer interaction. In Mittal et al., [19]. This volume.

25. T. Prater et.al. Kansas State robotics. In *Proceedings of the 14th National Conference on Artificial Intelligence*, July 1997, pg 797.

26. S. Shoval, J. Borenstein and Y. Koren. Mobile robot obstacle avoidance in a computerized travel aid for the blind. In *Proceedings of the 1994 IEEE Robotics and Automation Conference*, May 1994, pp 2023-2029.

27. R. Simpson, S.P. Levine, D.A. Bell, L.A. Jaros, Y. Koren, and J. Borenstein. NavChair: an assistive wheelchair navigation system with automatic adaptation In Mittal et al., [19]. This volume.

28. N. Tejima. Algorithm for eating noodle by a rehabilitation robot. http://www.ritsumei.ac.jp/bkc/ tejima/theme-e.html#noodle, Dept. of Robotics, Fac. of Science and Engineering, Ritumeikan Univ.

29. Wormald International Sensory Aids. 6140 Horseshoe Bar Rd., Loomis, CA 95650.

30. N. B. Zumel. A nonprehensile method for reliable parts orienting. Carnegie-Mellon University, Robotics Institute, Ph.D. thesis, January 1997.

Progress on the Deictically Controlled Wheelchair

Jill D. Crisman[1,2] and Michael E. Cleary[1]

[1] Robotics and Vision Systems Laboratory
Dept. of Electrical and Computer Engineering
Northeastern University
Boston, MA 02115
U.S.A.
[2] College of Computer Science
Northeastern University
Boston, MA 02115
U.S.A.

Abstract. We aim to develop a robot which can be commanded simply and accurately, especially by users with reduced mobility. Our shared control approach divides task responsibilities between the user (high level) and the robot (low level). A video interface shared between the user and robot enables the use of a deictic interface. The paper describes our progress toward this goal in several areas. A complete command set has been developed which uses minimal environmental features. Our video tracking algorithms have proven robust on the types of targets used by the commands. Specialized hardware and new tactile and acoustic sensors have been developed. These and other advances are discussed, as well as planned work.

1 Introduction

Our long term goal is to develop a "gopher" robot which could be commanded to navigate to a destination, retrieve objects, and return to a human operator. The robot will also be able to take the object back to a destination location. The navigation skills of this robot could also carry the person to the desired object. This device would be particularly useful as a navigation and reaching aid for those who use motorized wheelchairs and as a remote gopher robot for those who are otherwise mobility impaired. For example, consider a mobility impaired individual who has woken up early, before their health care professional has arrived for the day. He/she decides to read a book, but has forgotten their book in the other room. A robot that could be directed to the other room to retrieve the book would be extremely useful for this person.

One of two basic approaches are typically used for controlling robotic systems: teleoperation and autonomy. In the teleoperated approach, the human must provide either the desired position/velocity for each joint of the robot or the desired Cartesian space position and orientation for the tool of the robot.

V. O. Mittal et al. (Eds.): Assistive Technology and AI, LNAI 1458, pp. 137–149, 1998.
© Springer-Verlag Berlin Heidelberg 1998

To control a mobile robot, a joystick or similar device is frequently used to directly control the velocity and heading of the robot, as is commonly done in semi-autonomous control of wheelchair robots, e.g., [16,27,35,41]. An alternative approach allows the operator to use stereoscopically-presented images and a three-dimensional mouse to indicate way-points in the imaged space [39,40]. To control a tool, a six-dimensional mouse can be used to specify its desired Cartesian position and orientation. One alternative is to overlay graphics on a video view of the manipulator, allowing the operator to position and orient the tool as desired before commanding its motion, e.g., [3,20,24]. The advantage of the teleoperated approach is that the human has full control of all the degrees of freedom of the robot. In particular, if the human is controlling the robot in joint space, then singular conditions can be avoided. The difficulty with this approach is that the human must control many joints of the robot in real-time which can be difficult depending on the physical limitations of the individual and on the time allotted for the motion. It is also very difficult to control robots remotely via teleoperation since the orientation and motion of the robot do not match those of the operator and time delays further exacerbate the operator's intuitive sense of the robot's configuration [13].

The autonomous approach offers the most convenient interface for the human operator since the robot would interpret English-like commands such as "Get Moby Dick from the living room coffee table," e.g., [31]. However, the robot would need a large database of object models, or robust object recognition abilities which are beyond the state of the art [22,37] to perform a reasonable set of commands. To retrieve the book, for example, a model of the house is needed for the robot to know how to navigate to the living room coffee table. The robot would also need a model of "Moby Dick" as a book so that it can recognize it among other objects (including possibly other books) on the coffee table. Unfortunately a gopher robot using this scenario cannot retrieve any object for which it doesn't have a model. Although it is possible to interactively augment the robot's set of object/shape models, e.g., [17], and a variety of basic models could possibly be acquired from standard databases, this could be a time-consuming and tedious process since standard databases will not contain all the objects that the person may want the robot to retrieve. Even if the models are available, identifying objects from a large database of models can be very time-consuming for the robot. Another difficulty with autonomous systems surfaced in our conversations with users from the Disability Resource Center at Northeastern University. A number of people with disabilities, having had unsatisfactory or dangerous experiences with various new technologies on which they had to depend, are healthily skeptical of "autonomous" systems. As a result, they do not trust a robot to perform a task independently, and want to feel in complete control of their wheelchairs and aids. It may take a long time for this user community to accept and trust autonomous systems.

We are investigating a shared control paradigm for robotic systems which is easier for a human to control than a teleoperated system and does not require the object recognition techniques and large model databases of the autonomous approach. We view the human and the robot as working together to form the object retrieval/return system where the human performs the high-level recognition and planning tasks and the robot performs the low-level control tasks. For example, the human performs object recognition, route planning, and contingency planning while the robot performs safety shutoff and motion control. This is convenient from a system point of view since a user could more quickly and easily identify the book Moby Dick and plan a route to the living room coffee table than the robot could, for example. However, the human does not have to perform the tedious motion control of the joints. The robot performs local motion planning (via visual servoing [18]) and monitors safety sensors to stop any motions if unsafe conditions are encountered. Local motion planning uses a simple control loop that moves the robot based on the location of a few targets in the environment. The robot is able to quickly detect unsafe conditions simultaneously at many points around the robot (whereas the operator would constantly have to look around the robot to determine these conditions).

We use video images as the interface between human and robot since people are accustomed to identifying objects in video images (rather than acoustic images from sonar, for example) and since many researchers have been able to demonstrate real-time control of mobile robots using visual servoing, e.g., [10,11,14,19,34,38]. Therefore video images are convenient for both user and robot.

We call our robot *deictic*[1] [1] since the operator points out object features in the video images to the robot. The operator also tells the robot how to move relative to these features. The robot performs the local motion, then stops and waits for the operator to tell it the next local destination. In this way, the human must give a command to the robot every minute or so — much less frequently than in teleoperation. In addition, our robot does not need a model of either the environment or the objects on which the visual features occur since the robot is not performing the path planning or the object recognition. The system needs only form enough of a model of the object feature so that it can track the position of the target well enough to perform a visually servoed motion. Unlike other recent semi-autonomous, visually-guided mobile robot systems which allow the operator to point out navigational targets [19,34], our research has focused on developing a complete set of commands and targets to support *general-purpose* navigation. By general-purpose navigation we mean that the robot can go anywhere the operator desires and the robot is physically able to go, without requiring environmental modification or *a priori* world knowledge. Unlike the approach used by NASA's Sojourner lander [30,40] which allows way-points to be arbitrary points in 3-dimensional space, the deictic approach uses features of visible world objects as targets, providing the user a more direct link to the real world through which the robot is to move. The use of visual targeting also isolates the operator's command from the robot's motion in a way that joystick-like con-

---

[1] deic·tic (dīk'tik), *adj. Gram.* specifying identity or spatial or temporal location from the perspective of one or more of the participants in an act of speech or writing, in the context of either an external situation or the surrounding discourse. [32]

trol does not. This approach requires low bandwidth and is not time-critical. These characteristics are useful for robotic control by almost everyone [28], and should prove especially useful for users with disabilities which make real-time motor control difficult but who can still indicate targets, perhaps by use of an eye-tracking device, e.g., [15,33], or by voice commands, e.g., [3,23].

In this paper, we describe our progress at developing the gopher robot wheel-chair system. First, we have developed a complete command set for robotic navigation that uses corners and edges of objects as canonical object features and plans the robot motion relative to these object features in the environment. We have shown successful shared navigation through a large number of real and randomly generated environments using this small set of commands and object features. Second, we describe the video tracking system that is used to track the canonical object features for the robot visual servoing. We have successfully run this tracker on the canonical targets that we have used in our simulations. Next, we describe the hardware that we have developed for our experiments. Finally, we talk about our future plans for integrating and improving the components of this system.

2   Deictic Control Scenario

In our deictic control scenario, the user selects the type of motion that he/she wants the robot to perform and then he/she selects a target in a video image relative to which the robot will execute the command. In order for our shared control to be useful for the disabled community, we need to be sure that there is a minimal set of commands. This will avoid having the user weed through menu after menu before issuing a command.

We have discovered that a few commands can be used to navigate the robot through a wide variety of environments using a set of "canonical targets" [5,7]. Canonical targets are the parts of objects which are important for the robot's motion. We have found that the robot does not need to perform object recognition on the target, but instead needs only track features of objects which are important for its motions. For example, the extending corner of an obstacle to be avoided is a canonical target. The edge of a sidewalk or building is also a canonical target if the robot is following a path delineated by the sidewalk or building edge. The robot needs not identify the building or even the entire side-walk surface in order to navigate correctly. Specifically, we have discovered that we can direct our robot to navigate in almost any situation by identifying corners and edges. The video algorithm for tracking visual features on these canonical objects is given in the next section.

Commands to the robot have a verb description which describes the desired motion of the robot and the noun which describes the target relative to which the motion should be performed. The verb description has the following components: "motion," "direction," "target placement," and "distance or speed." These components are easily selected from a button based graphical user inter-face, but could just as easily be implemented by a sequence of selections which

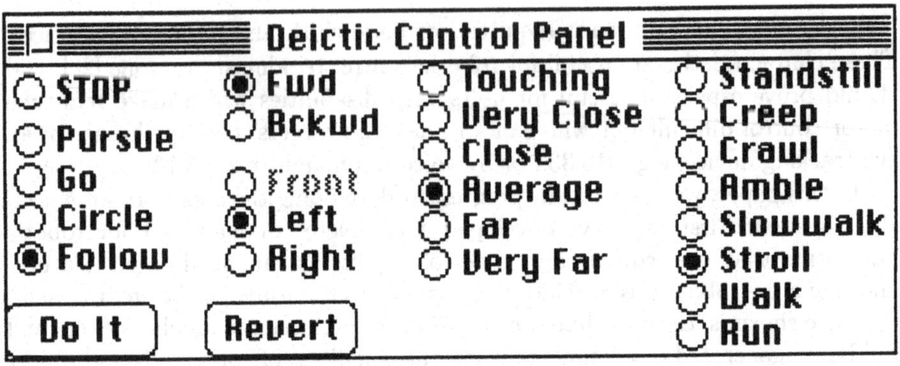

Fig. 1. Picture of the Deictic Control Panel for selecting the verb description.

could be accessed through switches. A picture of our button interface for the verb description is shown in Figure 1. First the user selects the motion: pursue (a moving target), go (to an target and then stop), circle (around an target) or follow (the edge of an object). Depending on the command, the user can then indicate on which side of the robot the target is to be located at the end of the motion. For example, if the user selects follow, he/she can then select left to say that the robot should move such that the target remains to its left. Next the user can select either the distance between the target and the robot or the speed at which the robot is to move (the other parameter will be computed from the one given). We have quantized the distances to 'touching', 'very close', 'close', 'average', 'far' and 'very far'. Similarly, we have quantized the speeds to 'standstill', 'creep', 'crawl', 'amble', 'slow walk', 'stroll', 'walk' and 'run'. Finally the user can indicate if he/she wants the robot to move forward (in the direction it is facing) or backward. A couple examples of a robot executing these commands in simulation are shown in Figure 2.

We have developed the deictic commands in simulation and have demonstrated successful cooperative navigation through a variety of environments accurately modeling the real world. We have also tested the commands in many randomly generated environments to ensure that we were considering all circumstances of navigation and have also navigated successfully through these environments [6]. In parallel, when simulating the robot's motion in models of real environments, we videotaped the motion of the robot's expected view of these environments to perform video tracking experiments. We describe our video tracking algorithm in the next section.

3 Progress on Canonical Video Target Tracking

The goal of canonical video target tracking is to track edges and corners as the robot is moving relative to those features. This implies that the adjacent surfaces will be changing their shape appearance in the video sequence. Therefore, we rely mostly on statistical color methods extended from previous work [10] to

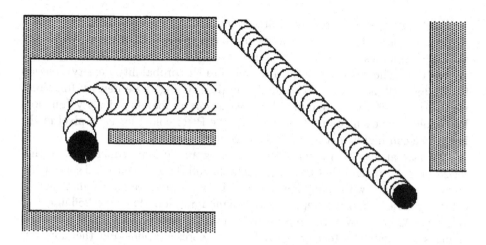

Fig. 2. Example of simulated robot executing a Circle closely with target on left (while moving forward) and a Go closely with target on left (while moving forward).

Fig. 3. Edge and Corner Model

"highlight" where the most likely canonical feature is located, following which we match simple models to these highlighted features.

The vision tracking can be thought of as containing two steps: training and tracking. The training uses the selected target from the user to form a color model of each object feature which will enable the system to quickly highlight object features in successive frames.

To specify a corner or an edge, the operator cursors two or three points to locate the edge in the image. These two edges should divide a small window into two distinctive surfaces as shown in Figure 3.

To form the shape model, a 64x64 window is drawn around the image location of the corner or edge and the two edges are extended or clipped to this window. All pixels on one side of this pair of lines form one region, R1, and all the pixels on the other side of the lines form the other region, R2 (Figure 3). A similar shape model can be formed for an edge where R1 and R2 would be separated by a line. To track surfaces, we have developed an algorithm that fits a deformable model to objects. However, we use color optical flow to better determine the motion of the surface object in the image [25].

Unlike other vision processing algorithms such as edge trackers, e.g., [2,36], we do not assume that each of the regions has a single surface color. Rather the algorithm assumes only that the distributions of colors are different in the two regions [10]. The system will quickly compute the probability density function for R1 and R2 as P(color|R1) and P(color|R2). Using these density functions, we use Bayes Rule to compute the *a posteriori* probability density function, P(R1|color) for each color. By computing the P(R1|color) for each pixel in the image, we can highlight the corner or edge in the image.

We have also worked extensively on finding an efficient reduction algorithm for 24-bit color images that best represents the full three-dimensional space (over 16 million colors) with many fewer colors. In particular, we have found that 52 color categories can better represent the 16 million colors than even 256 intensity values [12,42]. Moreover, this representation is much less sensitive to illumination changes than either intensity or RGB color representations and therefore our categorical color images are much more conducive for tracking object features outdoors.

We have demonstrated this tracking using variable-angle corners and edges. Although the video tracking can currently process 3 frames a second, we expect to get even faster processing when we optimize the algorithm. We have successfully demonstrated the tracking of a variety of difficult situations including tracking a corner of a wheelchair ramp where the two regions being tracked are of the same material and color [4]. In this case, the probability calculations enhance the difference in lighting of these two surfaces. We have also used straight edge corner models to track the edges of bushes and other natural objects. Our implementations of this algorithm have correctly controlled the pan and tilt of the camera system [8] and, more recently, have tracked door and table corner features while actively controlling the camera head on a moving robot.

4   Progress on Robotic Hardware

In a shared control robotic system, one needs actuators to physically perform the task, sensors to provide feedback on the status of the task being performed and a user interface to allow the human to communicate effectively with the robot. In this section we describe the hardware we are currently using to investigate the navigation and vision algorithms discussed in the previous sections, plus our work on alternative sensors.

The gopher robot must be able to navigate, reach out, and grasp objects. It may also need to position sensors to keep targets in view as the robot moves. For navigation, we have converted a motorized wheelchair to accept motion commands from a RS-232 interface. Figure 4 shows the wheelchair robot. We started with an Invacare Arrow wheelchair stripped completely except for the chair, motors, and batteries. We interfaced the wheelchair to a 386 PC-104. Motion controller cards from Motion Engineering provide the interface to the motors' power amplifiers. We installed optical encoders to measure the motion of the belt that runs from the drive to the wheels (therefore we can determine

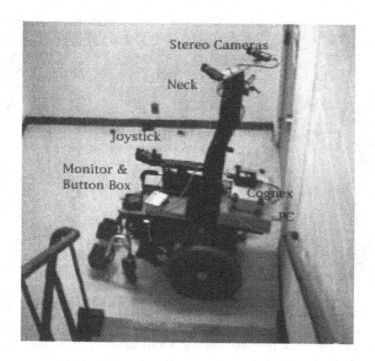

Fig. 4. Our Robotic Wheelchair

if the wheelchair is slipping). We have developed a "rack" that sits immediately behind the user that holds a box containing the controlling computer, a Cognex vision system (for the video and motion control software) and a Directed Perceptions pan-tilt unit which moves a pair of stereo cameras. The stereo cameras provide the main sensing of the system through a single digitizing board in the Cognex. An American Dynamics Quad Box combines the cameras' NTSC signals into a single video image which can be digitized as one. In this way, we ensure simultaneous acquisition of stereo images.

A Puma 200 is being used for reaching experiments. Currently our Puma is being retrofitted with a new interface to a standard 6-axis motor controller card. For a grasping tool, we have developed a new robotic hand, shown in Figure 5, which we call Graspar [9].

This hand has a tendoning system which allows the hand to mechanically conform its shape to the object being grasped. The advantage of this tool is that a wide variety of objects can be grasped without needing to compute the desired positions of each joint of the hand.

For safety, the robot must be able to detect when it bumps into objects or to insure that it is moving on flat surface. With the aid of Massa Products Corp. in Hingham, MA, we have developed a new directional ultrasonic transducer which has a fan-shaped beam rather than a conical shaped beam. This transducer gives our robot the ability to detect obstacles at a wide angle in one direction

Fig. 5. The Graspar Robot Hand. The left side of the figure shows Graspar mounted on a Puma 200 holding a screwdriver in a fingertip grasp. The right side shows a close up of the hand holding a soda can in an enclosing grasp from the top of the can.

and at a narrow angle in the opposing direction. This is particularly useful compared to the standard "ultrasonic rings" that are used in mobile robots. This transducer gives our robot the ability to see obstacles that are not at standard heights and give a higher resolution reading in the horizontal scan. A picture of the new transducer, its electronics package and a standard transducer and a representation of how this transducer could be mounted on a simple robot is shown in Figure 6.

While acoustic transducers are useful for detecting obstacles at a distance from the robot, they can still have problems with oblique object surfaces or highly reflective surfaces [26,28]. Therefore, we want to include active obstacle detection on our robot as well. Rather than whiskers that could miss a table top at a non-standard height, we designed a soft bumper similar in concept to others', e.g., [28], to be like thick "skin" that will cover our robot. To ensure that the robot can stop when moving at maximum speed, the bumper should be sensitive and thick enough to stop it before it damages itself or the environment. We developed such a sensor and tested it by mounting it on the hard footpad of the wheelchair. The bumper is soft foam with a thin piezo-electric film embedded in the foam as shown in Figure 7a. The film dramatically changes its electrical properties when it is bent or pressed. The foam was carefully selected to minimize the vibrations of the piezo-electric film when robot is moving. Filtering of the signal further reduces the electrical signals caused by vibrations. When large changes in this signal are detected, the robot is immediately stopped. We have

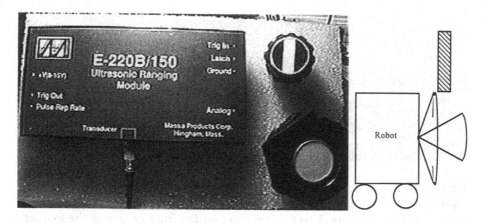

Fig. 6. New directional acoustic transducer produced by Massa Products Corp., Hingham, MA. The left shows the electronic control box for properly amplifying the return signal to be in the appropriate range for analog to digital conversion. The bottom right shows a standard Massa acoustic transducer in a mountable casing. The top right is the directional transducer. Notice that the ceramic surface of this transducer is masked to be linear rather than circular. The drawing on the right illustrates this transducer's usefulness for detecting overhanging obstacles compared to a standard conic sonar.

tested the bumper by mounting such sensors on the front of the foot pads of a wheelchair as shown in Figure 7b. We conducted experiments where we ran the wheelchair at full speed toward a wall. As the wheelchair collided with the wall, it stopped before the hard surface of the footpad touched the wall. This experiment was repeated many times and was successful in every case that we tried.

5   Conclusions and Continuing Work

We have made significant progress in the development of the control and real-time sensing algorithms for deictically controlled navigation on our wheelchair robot system. We have begun to implement these algorithms on the hardware, much of which we have developed in our lab. In this paper, we have described a wide variety of projects and experiments which we have performed that are directed to accomplishing the eventual system. We have also described some of the specialized hardware that we have developed and obtained for this project.

Currently, we are working on improving the speed of the target tracking algorithms and allowing more generally shaped corners to be tracked. We are also extending our color algorithms to adapt the color categories that are present in the scene. We are implementing the set of deictic commands we have developed for general-purpose navigation, including those which can be used with multiple targets (e.g., for docking operations) [6].

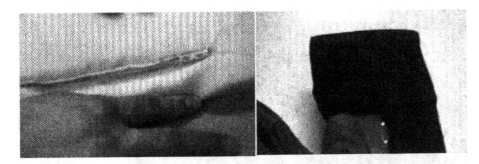

Fig. 7. Soft Bumper: On the left is the piezo-electric film embedded in firm foam. On the right is the finished bumper mounted on the footpad of the wheelchair robot. This bumper can detect any contact on the front or side of the footpad. In the future, we envision covering the wheelchair with similar bumpers.

6   Acknowledgments

This work is supported by the National Science foundation under grant number IRI-9210560 and under equipment grant number CDA-9320296. This work is also aided by the donation of a Cognex 4400 Machine Vision System by Cognex Corp., Needham, MA, and the donation of a Invacare Arrow motorized wheelchair from Invacare Corp. Some of the equipment for the conversion of the Arrow wheelchair was supported by the National Institutes of Health under a startup equipment grant for new biomedical research.

References

1. P.E. Agre and D. Chapman. What are plans for? *Robotics and Autonomous Systems*, 6(1,2):17–34, 1990.
2. S. Atiya and G.D. Hager. Real-time vision based robot localization. *IEEE Trans. Robotics and Automation*, 9(6):785–800, December 1993.
3. D.J. Cannon and G. Thomas. Virtual tools for supervisory and collaborative control of robots. *Presence*, 6(1):1–28, February 1997.
4. S.K. Chandwani. Feature tracking in video sequences using estimation. Master's thesis, Dept. of Electrical and Computer Engineering, Northeastern Univ., 1996.
5. M.E. Cleary and J.D. Crisman. Canonical targets for mobile robot control by deictic visual servoing. In *Proc., IEEE Int'l Conf. on Robotics and Automation (ICRA)* [21], pages 3093–3098.
6. M.E. Cleary. *Systematic Use of Deictic Commands for Mobile Robot Navigation*. PhD thesis, College of Computer Science, Northeastern Univ., September 1997.
7. J.D. Crisman. Deictic primitives for general purpose navigation. In J.D. Erickson, ed., *Intelligent Robotics in Field, Factory, Service and Space (CIRFFSS)*, pages 527–537. NASA and American Inst. of Aeronautics and Astronautics (AIAA), March 1994. Houston, TX. March 21-24.
8. J.D. Crisman and M.E. Cleary. Adaptive control of camera position for stereo vision. In D.J. Svetkoff, ed., *Optics, illumination, and image sensing for machine*

vision VIII, volume 2065, pages 34–45, Boston, MA, September 1993. SPIE – International Society for Optical Engineering.

9. J.D. Crisman, C. Kanojia and I. Zeid. Graspar: A flexible, easily controllable robotic hand. *IEEE Robotics and Automation Magazine*, 3(2):32–38, June 1996. U.S. Patent 80/197,384.

10. J.D. Crisman and C.E. Thorpe. SCARF: A color vision system that tracks roads and intersections. *IEEE Trans. Robotics and Automation*, 9(1):49–58, February 1993.

11. E.D. Dickmanns and V. Graefe. Dynamic monocular machine vision. *Machine Vision*, 1:223–240, 1988.

12. Y. Du and J.D. Crisman. A color projection for fast generic target tracking. In *IEEE/RSJ Int'l Conf. on Intelligent Robots and Systems (IROS)*, 1995.

13. W.R. Ferrell and T.B. Sheridan. Supervisory control of remote manipulation. *IEEE Spectrum*, 4(10):81–88, October 1967.

14. T. Fukuda, Y. Yokoyama, et al. Navigation system based on ceiling landmark recognition for autonomous mobile robot. In *Proc., IEEE Int'l Conf. on Robotics and Automation (ICRA)* [21], page 1720ff.

15. J. Gips. On building intelligence into EagleEyes. In Mittal et al. [29].

16. T. Gomi. The TAO project: intelligent wheelchairs for the handicapped. In Mittal et al. [29].

17. S. Hayati and J. Balaram. Supervisory telerobotics testbed for unstructured environments. *Journal of Robotic Systems*, 9(2):261–280, March 1992.

18. S. Hutchinson, G.D. Hager, and P.I. Corke. A tutorial on visual servo control. *IEEE Trans. Robotics and Automation*, 12(5):651–670, October 1996.

19. D.P. Huttenlocher, M.E. Leventon, and W.J. Rucklidge. Visually-guided navigation by comparing two-dimensional edge images. In *Proc., IEEE Conf. on Computer Vision and Pattern Recognition (CVPR)*, pages 842–847, 1994.

20. I. Ince, K. Bryant, and T. Brooks. Virtuality and reality: A video/graphics environment for telerobotics. In *Proc., IEEE Conf. Systems, Man and Cybernetics*, pages 1083–1089, 1991.

21. Institute of Electrical and Electronics Engineers. *Proc., IEEE Int'l Conf. on Robotics and Automation (ICRA)*, Minneapolis, MN, USA, 1996.

22. A.K. Jain and P.J. Flynn, eds. *Three-dimensional object recognition systems*. Elsevier Science Publishers (BV), North Holland, 1993.

23. Z. Kazi, S. Chen, M. Beitler, D. Chester, and R. Foulds. Multimodal HCI for robot control: towards an intelligent robotic assistant for people with disabilities. In Mittal et al. [29].

24. W.S. Kim and A.K. Bejczy. Demonstration of a high fidelity predictive/preview display technique for telerobotic servicing in space. *IEEE Trans. Robotics and Automation*, 9(5):698–702, October 1993.

25. J. Lai, J. Gauch, and J. Crisman. Computing optical flow in color image sequences. *Innovation and Technology in Biology and Medicine, A European Journal of Biomedical Engineering Promoting Application Research*, 15(3):374–390, 1994. Special Issue on Motion Computation and Analysis in Biomedical Imaging.

26. R.L. Madarasz, L.C. Heiny, R.F. Cromp, and N.M. Mazur. The design of an autonomous vehicle for the disabled. *IEEE Journal of Robotics and Automation*, 2(3):117–125, 1986.

27. D.P. Miller. Assistive robotics: an overview. In Mittal et al. [29].

28. D.P. Miller and M.G. Slack. Design and testing of a low-cost robotic wheelchair prototype. *Autonomous robots*, 2(1):77–88, 1995.

29. V. Mittal, H. Yanco, J. Aronis, and R. Simpson, eds. *Lecture Notes in AI: Assistive Technology and Artificial Intelligence*, 1998. This volume.

30. J.C. Morrison and T.T. Nguyen. On-board software for the Mars Pathfinder microrover. In *Second IAA Int'l Conf. on Low-Cost Planetary Missions*, John Hopkins Univ. Applied Physics Lab, Laurel, MD, April 1996. Int'l Academy of Astronautics (IAA).

31. N. Nilsson. A mobile automaton: An application of artificial intelligence techniques. In *Proc., Int'l Joint Conf. Artificial Intelligence (IJCAI)*, pages 509–520, 1969.

32. *Random House Compact Unabridged Dictionary*. Random House, 1996. Special Second Edition.

33. R. Rozdan and A. Kielar. Eye slaved pointing system for teleoperator control. In W.H. Chun and W.J. Wolfe, eds, *Mobile Robots V*, volume 1388, pages 361–371, Boston, MA, USA, November 1990. SPIE – International Society for Optical Engineering.

34. T. Shibata, Y. Matsumoto, T. Kuwahara, M. Inaba, and H. Inoue. Hyper Scooter: a mobile robot sharing visual information with a human. In *Proc., IEEE Int'l Conf. on Robotics and Automation (ICRA)*, pages 1074–1079, 1995.

35. R.C. Simpson, S.P. Levine, D.A. Bell, L.A. Jaros, Y. Koren, and J. Borenstein. NavChair: an assistive wheelchair navigation system with automatic adaptation. In Mittal et al. [29].

36. R.S. Stephens. Real-time 3D object tracking. *Image and Vision Computing*, 8(1):91–96, February 1990.

37. P. Suetens, P. Fua, and A.J. Hanson. Computational strategies for object recognition. *ACM Computing Surveys*, 24(1):5–61, March 1992.

38. C.E. Thorpe, ed. *Vision and navigation: the Carnegie Mellon Navlab*. Kluwer Academic Publishers, Boston, MA, 1990.

39. R. Volpe, J. Balaram, T. Ohm, and R. Ivlev. The Rocky 7 Mars rover prototype. In *IEEE/RSJ Int'l Conf. on Intelligent Robots and Systems (IROS)*, 1996. Also presented at Planetary Rover Technology and Systems Workshop, ICRA'96.

40. B.H. Wilcox. Robotic vehicles for planetary exploration. *Applied Intelligence*, 2(2):181–193, August 1992.

41. H.A. Yanco. Wheelesley, a robotic wheelchair system: indoor navigation and user interface. In Mittal et al. [29].

42. N. Zeng and J. Crisman. Categorical color projection for robot road following. In *Proc., IEEE Int'l Conf. on Robotics and Automation (ICRA)*, 1995.

Developing Intelligent Wheelchairs for the Handicapped

Takashi Gomi and Ann Griffith

Applied AI Systems, Inc. (AAI)
340 March Road, Suite 600
Kanata, Ontario
Canada K2K 2E4
{gomi,griffith}@applied-ai.com
URL: http://fox.nstn.ca/~ aai

Abstract. A brief survey of research in the development of autonomy in wheelchairs is presented and AAI's R&D to build a series of intelligent autonomous wheelchairs is discussed. A standardized autonomy management system that can be installed on readily available power chairs which have been well-engineered over the years has been developed and tested. A behavior-based approach was used to establish sufficient on-board autonomy at minimal cost and material usage, while achieving high efficiency, sufficient safety, transparency in appearance, and extendability. So far, the add-on system has been installed and tried on two common power wheelchair models. Initial results are highly encouraging.

1 Introduction

In recent years, the concept of applying behavior-based intelligent robots to service tasks [10] has been discussed. With the accelerated rate of aging of the population being reported in many post-industrial countries, demand for more robotic assistive systems for people with physical ailments or loss of mental control is expected to increase. This is a seemingly major application area of service robots in the near future. For the past six years, we have been developing a range of autonomous mobile robots and their software using the behavior-based approach [3,14]. In our experience the behavior-based approach [3,4,18,16,14] allows developers to generate robot motions which are more appropriate for use in assistive technology than traditional Cartesian intelligent robotic approaches [8]. In Cartesian robotics, on which most conventional approaches to intelligent robotics are based, "recognition" of the environment, followed by planning for the generation of motion sequence and calculation of kinematics and dynamics for each planned motion, occupy the center of both theoretical interest and practice. By adopting a behavior-based approach wheelchairs can be built which can operate daily in complex real-world environments with increased performance in efficiency, safety, and flexibility, and greatly reduced computational requirements.

V. O. Mittal et al. (Eds.): Assistive Technology and AI, LNAI 1458, pp. 150–178, 1998.
© Springer-Verlag Berlin Heidelberg 1998

In addition, improvements in the robustness and graceful degradation characteristics are expected from this approach. In the summer of 1995, an autonomy management system for a commercially available Canadian-made power wheelchair was successfully designed and implemented by our development team. The system looks after both longitudinal (forward and backward) and angular (left and right) movements of the chair. In addition, we implemented on-board capability to carry out "recognition" of the environment followed by limited vocal interactions with the user. The results were exhibited in August 1995 at the Intelligent Wheelchair Event organized by David Miller at the International Joint Conference on Artificial Intelligence (IJCAI'95) held in Montreal. Despite a very short development period (33 days), the chair performed remarkably well at the exhibition.

Encouraged by the initial success, we developed a three year plan to build a highly autonomous power wheelchair for use by people with various types and degrees of handicap. The intelligent wheelchair project, now called the TAO Project, intends to establish a methodology to design, implement, and test an effective add-on autonomy management system for use in conjunction with most common commercially available power wheelchairs. In order to demonstrate the principle, the project will build, during its life, an autonomy management system for several well-established electric wheelchair models currently available on the market throughout North America and Japan.

In late 1995, a sister R&D company was established in Japan exclusively for the development of intelligent robotic technologies for the disabled and the aged. With the initiative of this new R&D group, the development of TAO-2 autonomous wheelchair using a commercially available Japanese wheelchair began in the spring of 1996.

Based on our experience, methods used and some issues related to the application of the behavior-based approach to realize an intelligent wheelchair and possibly other assistive technologies are discussed. A brief survey is also presented of other groups who are working in this area.

2   Brief Survey of the Field

Below is a description of research on intelligent wheelchairs that has been conducted and still ongoing at some institutions. The survey is not intended to be complete but to provide an idea of the different approaches used.

2.1   IBM T.J. Watson Research Center

Some of the earliest work in the development of intelligent wheelchairs was a system implemented by Connell and Viola [6] in which a chair is mounted on top of a robot to make it mobile. Mr. Ed, as the chair was called, could be controlled by the user using a joystick mounted on the arm of the chair and connected to the robot. The user could also delegate control to the system itself to perform certain functions such as avoid obstacles or follow other moving objects. In addition to the joystick, input to the robot comes from bumper switches at the front and rear of the robot, eight infrared proximity sensors for local navigation and two sonar sensors at the front of the robot for   following objects. Control is passed from the user to the robot through a series of toggle switches.

A set of layered behaviors were used to control the chair's movement. These were broken into competencies with each small set of rules becoming a toolbox to achieve a particular goal. These groups could be enabled or disabled by means of switches controlled by the operator. It worked as a partnership in which the machine took care of the routine work and the user decided what needed to be done.

2.2   KISS Institute for Practical Robotics

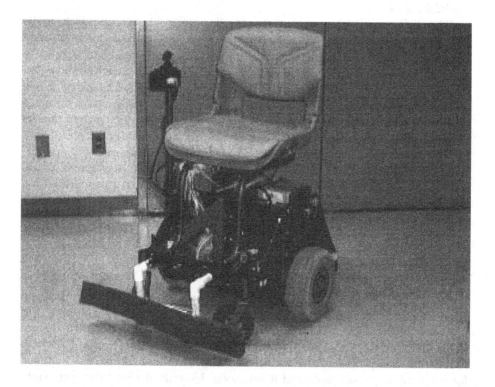

Fig. 1. TinMan II from KISS Institute

The KISS Institute for Practical Robotics (KIPR), located in Virginia is a non-profit educational corporation performing R&D on the integration of robotics in assistive technology, space robotics and autonomous underwater vehicles as well as education in robotics and related fields.

David Miller and Marc Slack at KISS Institute have developed TinMan I and II. In TinMan II shown in Figure 1, a supplementary wheelchair controller is installed between the joystick and the standard wheelchair motor controller. Along with sensors installed on the chair, the chair avoids obstacles and goes through openings with minimum input from the user. It has been tested with two power wheelchairs, Dynamics and Penny & Giles.

2.3   CALL Centre, University of Edinburgh

CALL Centre at the University of Edinburgh has developed the CALL Centre Smart Wheelchair. It was originally developed as a motivating educational and therapeutic resource for severely disabled children. The chairs were designed to assist in the assessment and development of physical, cognitive, social and communicative skills. Thirteen chairs have been built and evaluated in three local schools, one in a residential hospital and three others in pre-vocational establishments.

The chairs are adapted, computer-controlled power wheelchairs which can be driven by a number of methods such as switches, joysticks, laptop computers, and voice. The mechanical, electronic and software design are modular to simplify the addition of new functions, reduce the cost of individualized systems and create a modeless system. Since there are no modes and behaviors are combined transparent to the user, an explicit subsystem called the *Observer* was set up to report to the user what the system is doing. The *Observer* responds and reports its perceptions to the user via a speech synthesizer or input device.

The software runs on multiple 80C552 processors communicating via an I2C serial link monitoring the sensors and user commands. Objects or groups of objects form modules which encapsulate specific functional tasks. It is multitasking with each object defined as a separate task. The architecture of behaviors each performing a specific functional task is similar to Brooks' Subsumption Architecture.

2.4   University of Michigan

Simon Levine, Director of Physical Rehabilitation at the University of Michigan Hospital began development of NavChair in 1991 with a grant for a three year project from the Veteran's Administration [1,17]. The Vector Field Histogram (VFH) method was previously developed for avoiding obstacles in autonomous robots and was ported to the wheelchair. However, this method was designed for fully autonomous robots and it was soon determined that there were sufficient differences in the power base between robots and wheelchairs and in the requirements of human-machine systems that significant modifications were required. This resulted in a new method, called Minimum VFH (MVFH) which gives greater and more variable control to the user in manipulating the power wheelchair.

The NavChair (shown in Figure 2) has a control system designed to avoid obstacles, follow walls, and travel safely in cluttered environments. It is equipped with twelve ultrasonic sensors and an on-board computer. This team uses a shared-control system in which the user plans the route, does some navigation and indicates direction and speed of travel. The system does automatic wall following and overrides unsafe maneuvers with autonomous obstacle avoidance. Since it is desirable that the system change the user's commands as little as pos-

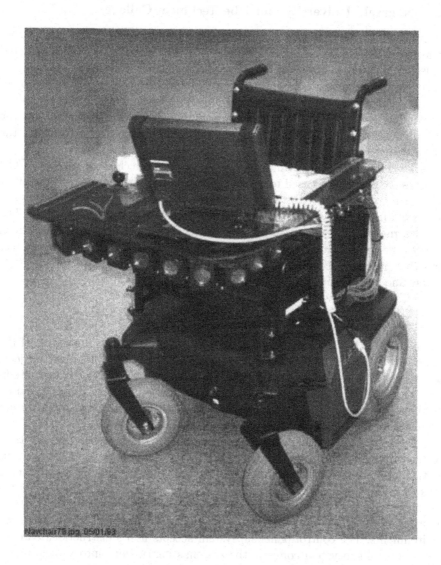

Fig. 2. NavChair, University of Michigan

sible, the system and user must cooperatively adapt to environmental or function conditions. A new method called "Stimulus Response Modeling" has been developed in which the system qualitatively monitors changes in the user's behavior and adapts in realtime. It is designed so that the adaptation is smooth and the change in modes intuitive to the user. By adjusting the degree of autonomy of obstacle avoidance the control modes of NavChair can be changed giving the user more or less control depending on the situation.

2.5   Nagasaki University and Ube Technical College

Existing ceiling lights in an indoor environment are used as landmarks for self-localization of a motorized wheelchair by [19]. The chair is therefore restricted to use within one building, the layout of which is known in advance. An azimuth sensor is used to give the angle between a fixed point and a particular object and a vision sensor detects the ceiling lights. The ceiling lights are used as the landmarks but if the lights are missed then the azimuth sensor and the rotating angle of both wheels provide the information necessary to continue the navigation.

A laser range finder is used to detect obstacles in the chair's path. Two CCD cameras are used, one is used to detect the ceiling light landmarks and the other is used in conjunction with the laser range finder to detect objects. A slit-ray is emitted from the laser emitter and this is detected by the CCD camera. The image signal is processed by a logic circuit constructed with an FPGA which informs the controller if passage is clear or where obstacles exist. In twenty test runs in a room with ten ceiling lights the maximum position error was 0.35 meters and the maximum orientation error was 17 degrees.

2.6   TIDE Programme

Technology initiative for disabled and elderly people (TIDE) programme of the European Union began in 1991 as a pilot action with 21 development projects and a budget of ECU18 million. The SENARIO project (SENsor Aided intelligent wheelchair navigatIOn), one of the initial projects within TIDE, includes 6 member companies from Greece, Germany, the UK, and France to introduce intelligence to the navigation system of powered wheelchairs.

The system consists of five subsystems: risk avoidance, sensoring, positioning, control panel, and power control. The risk avoidance subsystem includes the central intelligence and inputs information from the sensoring and positioning subsystems. The sensoring subsystem includes ultrasonic, odometer, and inclinometer sensors. The positioning subsystem identifies the initial position of the chair by means of a laser range finder and allows the chair to be used in known environments. The control panel subsystem accepts user's instructions and the power control subsystem converts the system's instructions into vehicle movements.

The system has two modes of operation, the Teach mode and Run mode. In the Teach mode the user selects the desired paths from a topological diagram. In the Run mode (on a predefined path) the user selects a path and the system will follow it based on stored information obtained during the Teach mode. On a free route, the system takes instructions from the user and navigates semi-autonomously while monitoring safety and taking action or warning the user of the level of risk.

2.7   Wellesley College, MIT

Wheelesley is the name given to the chair used for experimental development by Holly Yanco, first at Wellesley College and now at MIT [21,20]. This chair has a Subsumption Architecture-like layered approach to its performance. By means

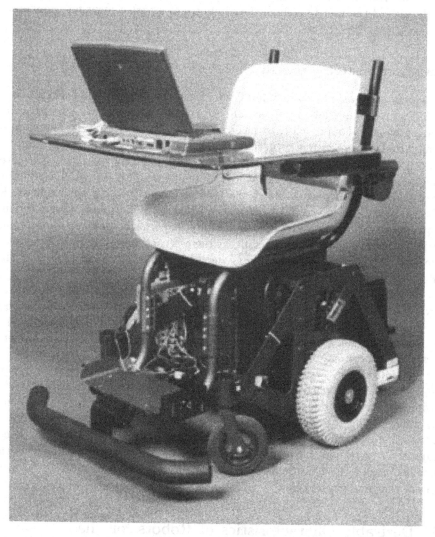

Fig. 3. Wheelesley Robot

of a graphical interface the user of the chair points to the direction in which the chair should head. The chair then goes in that direction while performing other tasks such as obstacle avoidance. The interface also allows the user to tell the chair when specific tasks such as going up a ramp are required and to have a record of a particular environment and important features of that environment.

The chair is designed in such a way that it can turn in place. It has 12 proximity sensors, 6 ultrasonic range sensors, 2 shaft encoders and a front bumper with sensors. A 68332 computer is onboard and the interface runs on a Macintosh Powerbook. Work is underway to incorporate information from the angle of the eyes of the user to control the computer as a replacement for the mouse.

2.8   Northeastern University

The long-term goal of Crisman and Cleary [7] is to develop a robot which can go to a destination, retrieve an object and return it to the operator. A teleoperated and autonomous approach each has its strength and weaknesses. Therefore, a shared control approach is suggested to divide the task between the user and the robot, taking advantage of the strengths of each. The user performs high-level functions such as object recognition and route planning while the robot performs safety and motion controls. Since the user points the objects out explicitly in a video image, the robot has been named "Deictic." The robot, after receiving instructions how to move relative to the object, performs the local motion and waits for further instruction. This means there is continuous interaction between the user and the robot with the user giving instructions to the robot every minute or so.

Commands are given to the robot by means of a button interface in which a verb description describes the desired motion of the robot and a noun describes the object relative to which the motion should be performed. The robot is able to navigate in almost any situation using its vision system to identify corners, edges, and polygonal patches.

The initial work was done in simulation followed by an implementation on an Invacare Arrow wheelchair. Motion controller cards, optical encoders, and a vision system were added to the wheelchair. New directional ultrasonic transducers were developed to detect obstacles at a wide angle in one direction and at a narrow angle in the opposite direction. This gave the robot the ability to detect objects not at standard height. A bumper with piezo-electric film embedded was installed to detect when the chair did bump an obstacle. A Puma 200 was used for the reaching experiments.

3   Desirable Characteristics of Robots for the Handicapped

3.1   Background

Since around 1992, AAI began a number of exchanges with people with various handicaps and the individuals who assist them. This was preceded by a few years of on-going interactions with the handicapped community through marketing, installing, servicing, and training individuals on a speech-to-text voice interface system for computers. This device proved to be effective for people with

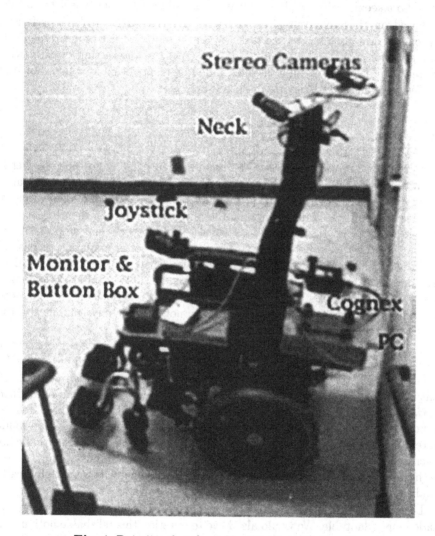

Fig. 4. Deictic robot from Northeastern University.

several types of handicap, particularly for individuals who had lost arm/hand usage. Since late 1995, voluntary work has been attempted by members of AAI at two institutions for the mobility handicapped in Japan: a senior citizen's hospice for severe physical/mental problems, and an institution for people with severe physical handicaps. A considerable amount of time practising physical assistive work has been carried out by members of the R&D team, including the designer involved in the conceptual design of the robots, engineers and a technician responsible for the construction of the robots, and the project manager and administrators of the robotics projects. In early 1995, an individual with a severe physical disability (a quadriplegic) joined AAI as a regular data entry/bookkeeping clerk and as a future tester of autonomous wheelchairs.

Based on these exposures, as well as earlier volunteer work, a preferable approach to robotics for service tasks [9] and a tentative list of desirable characteristics for future robots built for the purpose of interacting directly with severely handicapped or fully disabled individuals has been compiled. Some of the desirable characteristics are discussed below.

3.2 Softness and Flexibility

Establishment of rapport between the handicapped person and the caregiver is essential for the care to be successful. So much so, there will be a great deal of anxiety in those treated by future robotized arms, support boards, and wheels. The need for softness realized between the physical interface of the end effectors of such a robot and the human body surface or limbs does not stop at simple padding of otherwise solid effector surfaces, or use of softer materials, or passive or active compliance of effectors. The softness must also be architectural in that the entire physical support structure must be able to alter, reconfigure, and even completely restructure moment to moment reactions and responses to accommodate, whenever necessary, changes in not only the physical but also the perceived psychological situation of the user.

The flexibility of the system as a whole, as well as that of the end effectors, must essentially come from this "structural softness." The flexibility must be founded on the openness of the design of motions the system can generate so that it does not rely on fixed modes of operation or rigid scenarios defined *a priori*. In most circumstances humans in general behave without a prepared set of motion patterns, and since we are dealing with such an existence, a man-made system itself must not act with a fixed set of motions which are algorithmically describable. This places the appropriateness of most existing system control methods in doubt as a tool to seriously deal with many types of physically handicapped people.

Learning has often been hailed as a scheme with which a system can be made more adaptable. We would also have to question this relishable notion as a candidate that would sufficiently increase adaptability of systems such as service robots dealing directly with humans. Learning schemes, particularly those so far studied to the greatest extent and depth in the symbolic AI community, have failed to make significant contributions to robotic systems operating in highly dynamic application areas. In general, learning research has focussed on methods to improve the chosen performance index of systems but variables involved in the scheme are most often not grounded through sensors or actuators.

3.3   Fail Safe and Robust

A robot arm holding a fragile human body must not drop the person when a bug is hit for the first time. The concept of fail safe implies readiness of a system against possible failure. In traditional system engineering disciplines, such as Fault Tolerant Computer Systems (FTCS) research and practice, this typically translates into the preparation of additional capabilities in the form of a standby in computer hardware and software. The concepts of hot-standby and cold-standby are commonly employed in system design. Since it is impossible to prepare for every possible failure, the provision of readiness should exist, however, more in the form of capabilities spread across the system in atomic form and meshed fine grain with the competence structure which also functions in the normal execution of tasks. This is analogous to the way readiness to failure is implemented in life forms found in nature. If a small animal or an insect temporarily loses the use of a limb, it tries to adjust to the situation by immediately enlisting the use of other limbs or even other portions of the body. The additional capability readied in this form would be quickly organized and mobilized the moment a fault is detected.

3.4   Graceful Degradation

A cousin to the concept of fail safe, graceful degradation is more important in systems that physically interface with humans than in systems that deal with materials and artifacts. A control system designed as a monolith or components with relatively larger granularity would have less chance of realizing the concept fully. When one loses a limb, the resulting transition is not smooth, causing great suffering to the individual. However, every day we lose a large number of brain cells that we know won't reproduce, but we do not deteriorate or lose capabilities as drastic as loosing a limb. Systems composed of finer grain active units seem to offer more desirable results.

3.5   Evolvability

Another reason for the failure of learning in symbolic AI would be the relatively short time the methods have typically tried to achieve the "result." In fact, we probably do not know what desirable results are as much as we think we do. Both shortcomings, this and the lack of grounding, are due mostly to the very nature of being symbolic rather than pragmatic.

In evolution, changes occur along a much longer time scale. In *situated* and *embodied* systems, such as life forms in nature and well-built autonomous robots, a search through a very high dimensional space of the real world for adaptation demands "experiments" on a vast number of combinations of dimensional parameters, if such dimensionalization or parameterization makes sense at all. Evolutionary Robotics (ER) is an emerging field of science and technology [12], where physical or virtual robots' autonomy structures are evolved to achieve collective trans-generational learning. ER seems to be a scheme that could well be applied to robots operating to tend and care for humans because of the open nature of human autonomy and ER's basic principle that can provide long term learning. Here, the concept of learning should probably be replaced by a more comprehensive concept of evolution, which implies perpetual adaptation of an autonomous system to a constantly changing operational environment rather than optimization of one or more performance indices of such a system.

3.6   The Development Plan

The development of autonomous wheelchairs at AAI is carried out in the following four phases. Some of the phases overlap in their execution.

1. The basic safety phase,
2. The mobility phase,
3. The human interface phase, and
4. The exploration phase.

Currently, we are in the second phase of the project which began on April 1, 1996. Prior to the start of the project on July 20, 1995, a study was conducted to identify various requirements by potential users of the autonomous wheelchair both in Canada and Japan through interactions with people with various types of handicap. Causes of the handicaps we came across included gradual mobility loss by aging, recent sudden loss of body control due to brain damage, and prolonged motion limitations and bodily contortion due to stroke suffered at a young age. The project continues to enjoy cooperation from institutions for the handicapped and individuals with disabilities. The TAO project is scheduled to end in the summer of 1998. For a description of the development plan, please refer to [11].

4   Implementation of the First Prototype, TAO-1

A regular battery powered wheelchair (a motorized chair) produced and marketed in Canada (FORTRESS Model 760V) was used as the base of the first implementation of the concept. A set of sensors, a computerized autonomy management unit, and necessary harnesses were built and added to TAO-1 (Figure 5) through the summer of 1995.

4.1   Planned Functions of the Chair

The selection of functions to be implemented on TAO-1 was somewhat influenced by the rules set out for the IJCAI'95 robotics contest. However, later demonstrations of our prototype and observations made at an institution for the aged confirmed that the guideline was in fact appropriate. Of the following functions which we now follow, only the first two were attempted at our IJCAI'95 entry. However, all five of them are currently pursued.

(a) **Basic collision avoidance** This is achieved by behaviors which monitor and respond to inputs from on-board CCD cameras or those which respond to active infrared (IR) sensors. When the chair encounters an obstacle, it first reduces its speed, and then depending on the situation it faces, stops or turns away from the obstacle to avoid hitting it. The obstacle can be inanimate (e.g., a column in a hallway, a light pole on the sidewalk, a desk, a standing human) or animate (a passerby, a suddenly opened door in its path, an approaching wheelchair). Encountering a moving obstacle, the chair first tries to steer around it. If it cannot, it stops and backs off if the speed of the advancing obstacle is slow enough (e.g., 20 centimeters per second). Otherwise, it stays put until the obstacle passes away. Thus, if the chair encounters another wheelchair, both chairs can pass each other smoothly as long as there is enough space in the passage for two chairs. A fast paced human usually does not affect the chair's progress and at most causes the chair to temporarily slow down or steer away.

(b) **Passage through a narrow corridor** When surrounded by walls on each side of the path, as in a hallway, the chair travels autonomously from one end to the other parallel to the walls.

(c) **Entry through a narrow doorway** The chair automatically reduces its speed and cautiously passes through a narrow doorway which may leave only a few centimeters of space on each side of the chair. Some types of ailment such as Parkinson's disease or polio often deprive a human of the ability to adjust the joystick of a power wheelchair through such a tight passage.

(d) **Maneuver in a tight corner** Similarly, when the chair is surrounded by obstacles (e.g., walls, doors, humans), it is often difficult to handle the situation manually. The autonomous chair should try to find a break in the surroundings and escape the confinement by itself unless instructed otherwise by the user.

(e) **Landmark-based navigation** Two CCD color cameras on-board the chair are used for functions explained in (a), (b), and (c) above. They constantly detect the depth and size of free space ahead of the chair. The cameras are also used to identify landmarks in the environment so that the chair can travel from its present location to a given destination by tracing them. An on-board topological map is used to describe the system of landmarks.

4.2   Hardware Structure

As a standard powered wheelchair, model 760V has two differentially driven wheels and two free front casters. Although they are designed to rotate freely around their vertical and horizontal axis, these casters typically give fluctuations in delicate maneuvers due to mechanical hysteresis that exists in them because of design constraints (the rotating vertical shaft of the support structure of the caster cannot be at the horizontal center of the caster). This sometimes causes the chair to wiggle particularly when its orientation needs to be adjusted finely.

Fig. 5. Autonomous wheelchair TAO-1

Such fine adjustments are necessary typically when a wheelchair tries to enter a narrow opening such as a doorway.

The entire mechanical and electrical structure, the electronics, and the control circuitry of the original power wheelchair were used without modification. The prototype autonomy management system still allows the chair to operate as a standard manually controlled electric wheelchair using the joystick. The joystick can be used anytime to seamlessly override the control whenever the user wishes even in autonomy mode.

Physical additions to the chair were also kept to a minimum. AI components added to the chair were made visually as transparent as possible. Two processor boxes, one for vision-based behavior generation and the other for non-vision behavior generation are tacked neatly under the chair's seat, hidden completely by the wheelchair's original plastic cover. Sensors are hidden under the footrests, inside the battery case, and on other supporting structures. Only the two CCD cameras are a little more visible: they are attached to the front end of the two armrests for a good line of sight. A small keypad and miniature television set are installed temporarily over the left armrest to enter instructions and for monitoring.

The non-vision behavior generator is based on a Motorola 68332 32-bit micro controller. A multi-tasking, real-time operating system was developed and installed as the software framework. This combination gave the system the capability to receive real-time signals from a large number of sensors and to send drive outputs to the two motors which govern the wheels. The chair currently has several bump sensors and 12 active infrared (IR) sensors which detect obstacles in close vicinity (less than 1 meter) of the chair. Signals from the cameras are processed by a vision-based behavior generation unit based on a DSP board developed by a group at MIT. Vision processing is discussed in Section 6.6 below.

4.3 Software Structure

The over-all behavior structure of TAO-1 is shown in Figure 6. Smaller behaviors are lumped up to save space on the diagram. Software for the vision system is also built according to behavior-based principles. The major difference between this and conventional image processing is that it consists of behaviors, each of which generates actual behavior output to the motors. It can presently detect depth and size of free space, vanishing point, indoor landmarks, and simple motions up to 10 meters ahead in its path. Indoor landmarks are a segment of ordinary office scenery that naturally comes in view of the cameras. No special markings are placed in the environment for navigation.

There are also a large number of behaviors invoked by IRs and bumpers which collectively generate finer interactions with the environment. Vision-based and non-vision behaviors jointly allow the chair to proceed cautiously but efficiently through complex office spaces. Note that there is no main program to coordinate behaviors.

Currently, the autonomy program occupies about 35 KBytes for all of the vision related processing and 32 KBytes for other behavior generation and miscellaneous computation. Of the 35 KBytes for vision related processing, only about 10 KBytes are directly related to behavior generation. The rest are involved in various forms of signal preprocessing: generation of depth map, calculation of the size of free space, estimation of the vanishing point, and detection of specific obstacles in the immediate front of the chair.

Wheelchair Behavior Structure

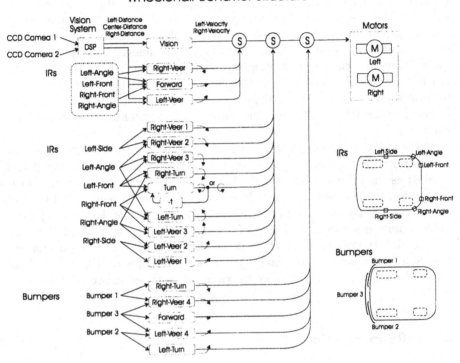

Fig. 6. TAO-1 behavior structure (not all behaviors are shown).

Of the remaining 25 KBytes, approximately 20 KBytes are used in the neural network system for detecting landmarks and referencing a topological map. The current implementation of the landmark system consumes only 256 Bytes per landmark, although this figure may change in the future as more sophisticated landmark description might become necessary. The current system has space for up to 64 landmarks but this can also be adjusted in future versions.

Of the 32 KBytes of non-vision processing (i.e., processing of inputs from IR's , bump sensors, voice I/O, etc.), again no more than several KBytes are spent for generating behaviors. Altogether, there are some 150 behaviors in the current version of TAO-1. A considerable amount of code has been written to deal with trivial periphery, such as keypad interface, voice I/O, and LCD display. The comparable inefficiency of coding is because these non-behavioral processing had to be described in more conventional algorithms.

5   The Second Prototype, TAO-2

Encouraged by the success of TAO-1, in late 1995 a sister company of AAI (AAI Japan, Inc.) was established in northern Japan. AAI Japan is dedicated to the development of advanced intelligent robotics to aid people with various handicaps. In May 1996, AAI Japan purchased a new power wheelchair (Suzuki MC-13P), which is a model widely used in Japan. MC-13P has a form of power steering in which the two front casters alter their orientation in synchrony with the drive wheels when a turn is indicated by the joystick. The servo controller also halts the inside turn wheel of the two drive wheels while the chair is making a tight turn. This is a significant departure from the way the FORTRESS model makes a turn. The latter simply turns the two differentially driven main wheels in opposite directions, allowing the chair to turn in place. The intent of providing a power steering feature on the Suzuki chair is obviously for ease of use, and the user is freed from the wiggly caster problem described above. However, this prevented the chair from making turns in a tight turn circle. The feature was felt undesirable for an autonomous chair.

Immediately following the purchase of the Suzuki chair, the development team began building an autonomy management system for TAO-2; a new prototype autonomous chair based on MC-13P. The over-all computer hardware and software structures as well as sensors are almost identical to those for TAO-1, except for a few changes listed below to accommodate the above mentioned and other minor differences in characteristics.

1. The behaviors responsible for turning TAO-2 needed their parameters adjusted.
2. The locations of touch sensors made up of thin piano wires needed to be moved forward in order to compensate for a larger turn circle.
3. The back bumper was not activated since it was hardly used. The difference in turning characteristics reduced the chance of the Suzuki chair performing frequent switch backs.
4. Two prominent side bumpers were added to protect bystanders when the chair makes a turn in their direction. This was necessitated by the lack of structure on which to mount sensors.

TAO-2 is shown in Figure 7. It was fitted with the autonomy management system at AAI in Canada in the span of one week. After two days of testing, it was shipped back to Japan in time for a public demonstration in the town of Kosaka, Akita Prefecture.

Fig. 7. TAO-2 autonomous wheelchair

6 Evaluation of the Prototypes

6.1 Demonstrations

When TAO-1 was demonstrated at IJCAI'95 in Montreal on the 22nd of August, it was the 33rd day of the development of the first prototype. Everything from the motherboard, vision system, sensor arrangements and their harnessing, operating system (based on an earlier prototype), a large number of behaviors (some 60 by that time) were all developed and tested in that period. The chair could perform functions (a) and (b) in Section 4.1 well and functions (c) and (d) moderately well, although they were not initially targeted. Function (e) was not yet implemented. In all, it performed as well as other chairs at the exhibition most of which took much longer time to develop. All five functions are now implemented on TAO-1 and are undergoing continuous improvement.

TAO-2 was demonstrated on June 4, 1996 at a gymnasium of a local school in Kosaka, Japan. The chair ran smoothly throughout the 1 hour demonstration persistently avoiding by-bystanders, other obstacles and the walls. Unsolicited, a severely handicapped spectator who could not even reach the joystick volunteered to test ride the chair. The chair performed to her satisfaction and excitement as it went through the gymnasium among a large number of spectators.

The success of the two prototypes suggests that our intention to build a standardized add-on autonomy unit is a valid one. The concept has at least been proven in two power wheelchair types which come from drastically different backgrounds. The divergence in design philosophy and practical variances in implementation, some fairly significant, of a base power wheelchair can be absorbed by relatively minor hardware and software alterations made on the standardized add-on unit. TAO-2 also showed that the installation, testing, and adjustment of a separately built autonomy unit can be made in a very short period of time. In both TAO-1 and TAO-2, no cooperation from the manufacturers was sought. In each case, characteristics of the joystick were studied and a seamless interface was designed around it.

6.2   TAO-2 Experiments

After successfully testing the basic navigation functions of TAO-2 at our laboratory in Canada, it was transported to AAI Japan's facility in Akita Prefecture, Japan in May, 1996 for additional tests and adjustments. Two types of experiments were conducted with TAO-2: indoor experiments and running of the autonomous chair outdoors in snow. The indoor experiments included unassisted navigation of the chair in a circular corridor and the gymnasium of a local primary school, and in corridors of an institution for physically handicapped adults. At the school, the chair navigated smoothly both in the circular corridor and the gymnasium except when it hit a glass door separating the corridor and one of classrooms next to the corridor. The incident was due to the fact that the chair bases its collision avoidance on vision (incapable when faced with a planer glass surface under rare lighting conditions) and active infrared (IR) sensors (IR emission is transparent through most glass surfaces). This, however, does not mean the present sensors are inferior. On the contrary, combined they are vastly more efficient and capable than other sensors such as laser range finders and ultrasonic sensors. Nevertheless, the addition of local ultrasonic sensors is being considered to cover this imperfection.

In the gymnasium which was populated by several dozen spectators, some of whom were surrounding the chair, TAO-2 constantly found a break in the crowd and escaped from the human wall without touching anyone. A female spectator with severe bodily contortion volunteered to try the chair. Her condition was such that she was not even capable of extending her arm to reach the joystick. As in TAO-1, the control structure of the original power wheelchair (Suzuki MC-13P model) was left intact when the autonomy management system was added. The intelligent chair is designed to allow the user to take over the entire control system by touching the joystick. It then simply acts as a standard motorized chair. Despite the total absence of input from the user, the chair navigated smoothly, always successfully avoiding walls and spectators. When completely surrounded by the spectators, it stopped until a break which was approximately 50% wider than the width of the chair developed roughly in front of it. It then moved out of the circle through the opening. The ability to locate a break anywhere in a circle regardless of its orientation when surrounded by people has been implemented and tested in other behavior-based robots.

When tested at a local institution for the severely physically handicapped, the chair managed to travel along corridors in most cases. Interest in an autonomous wheelchair that can take individuals to a desired destination was strong, and the experiment had to be conducted amid many spectators who were themselves in a chair. TAO-2 encountered some difficulties when surrounded by other wheel-chairs in close proximity. This difficulty includes at its core a common problem for both TAO chairs: the autonomy management system still requires better processes to detect thin pipes or tubes in the environment. Such processes will likely depend on inputs from the vision system as it provides the widest com-munication path between the chair and the environment and is amenable to the addition of new processes to deal with specific problems such as detection of ver-tical and horizontal thin pipes in the path of the autonomous chair. Landmark navigation was not attempted in these experiments due to the shortage of time and manpower necessary to prepare an on-board topological map. In all, TAO-2 at this stage appeared to have basic navigational capacity in populated indoor space.

In February 1997, TAO-2 was tested outdoors on the snow covered pavement and sidewalks of Kosaka, Japan. No particular modifications were made to the basic functioning of the indoor version of the chair except for minor adjust-ments to the vision system, the active IR sensors and the software. The outdoor temperature was around −10 degrees Celsius when the chair was tested. First, TAO-2's ability to interpret signals obtained through the vision system and other sensors (IR's and bumpers) when navigating through the mostly white surround-ing snow-scape was checked. The chair successfully navigated through a narrow corridor sided by walls of snow. Most of the time the chair depended on both the vision system and IR sensors to position itself roughly in the middle of the narrow (changing from 1.2 to 1.5 meters) corridor. The surface of the floor of the corridor was mostly covered by snow with some foot prints. The height of the snow walls on both sides of the corridor was about one meter. The sunlight which was shining through a thin layer of clouds at an angle from behind the chair caused one of the walls to appear quite dark and the other slightly brighter, while the floor was yet another tone. Such a contrast was good enough for the vision system to distinguish the geometry and guide TAO-2 roughly in the mid-dle of the snow corridor. Whenever the chair's course noticeable deviated from the center of the corridor, mostly due to friction and slippage caused by the uneven surface of the snow covered floor, the IRs on either side would detect the deviation and associated processes were invoked to cancel the deviation.

When TAO-2 traveled through the entire length of the corridor and reached the open pavement which was mostly covered by snow with some tire marks and sporadic black exposed surfaces of asphalt, it navigated among these ground marks just as humans would try to make sense of the orientation of the hidden roadway underneath the largely snow-covered pavement.

The TAO-2 chair was also tested on a sidewalk under similar climatic condition (snow on the ground, cloudy day with sufficient light, −10 degrees Celsius). However, the surface of the sidewalk was clear of snow because of a snow removal system that warms up the underside of the surface of the sidewalk using well-water. The system very successfully maintains a snow-free strip about 90 centimeters wide in the middle of a 1.2 meter wide sidewalk up until a certain temperature and rate of snowing. This optical contrast created an ideal condition for the vision system. Because of the high contrast between the wet surface of the sidewalk made up of dark brown bricks of the sidewalk and the white snow covered edges of the sidewalk, the vision system could easily follow the track. Light standards are erected at regular intervals on the edge of the sidewalk creating a particularly narrow passage. When passing by the light standards, the chair slowed down to negotiate past them but did not have particular difficulties to clear them. In general, the performance of TAO-2 in snowy outdoors was much better than expected. It became clear that the chair can cover the basic navigational requirements through a snow-covered town where a distinctive sidewalk system with snow removal is available.

6.3   Development Time

The extremely short development time required for the initial prototype for both TAO-1 and TAO-2 can largely be attributed to the behavior-based approach. To achieve the demonstrated level of mobility and flexibility would normally have required another several months to a few years in conventional AI-based mobile robotics. In behavior-based robotics, the operational characteristics of the sensors need not be as precisely uniform as in conventional mobile robotics. For example, emission strength and angular coverage of the emitter, and the sensitivity and shape of the reception cone of the receptor of on-board IR sensors need not be homogeneous across all sensors, allowing the use of inexpensive sensors and simpler testing.

All sensors, including the CCD cameras, need not be installed at precise translational and angular coordinates. They also do not need calibration. They were placed on the chair in a relatively *ad hoc* manner at first, and continually moved around for better results as the development went on. In fact, the cameras and some of the sensors are attached to the chair by velcro detachable tape, so that their location and orientation can be adjusted easily. Such loose treatment of sensors is not common in conventional robotics where the robot's motions are derived after high-precision measurements of the relationships between its extremities and the environment. The large tolerance for signal fluctuation is due also to flexibility of processing and greater adaptability inherent in Subsumption Architecture [3].

With the absence of "sensor fusion," sensor inputs are directly linked to motor output only with a simple signal transformation and amplification (e.g., from sensor output voltage to motor drive current). The developer only needs to adjust the appropriateness of the definition and performance of the sensor-action pair or behavior in terms of its output without a detailed and precise analysis of input signal characteristics and elaborate planning and computation of output signal generation. Readers not familiar with the theoretical basis of behavior-based AI are encouraged to read [5]. These theories are fully put into practice in our development.

6.4   Software Structure

During development, sensor-actuator pairs or behaviors are simply "stacked up." They are added to the system one by one without much consideration for the design of the over-all software structure. Our operating system provided an adequate framework for the incremental development process allowing for shorter development time.

Thus, software development went totally incrementally side by side with finer adjustment of the sensors. Only general functions needed to be assigned to each sensor-actuator pair type first. For example, depth map - motor pairs are excellent for dealing with obstacles that suddenly appear in the path of the chair a few meters away. But the same sensor-actuator pair type is not at all effective for the management of the situation in which the chair has actually made physical contact with an obstacle.

Sometimes, competition or contradiction occurs between two or more behaviors. Such contradicting definitions of behaviors are in most cases easily observable and corrected quickly. An example of more complex contradiction occurs when two IR collision-detection sensors placed on the left and right front sides of the chair detect an approaching doorway in quick succession. Since the doorway is normally quite narrow, the reflection of infrared signals received by these sensors is usually strong enough to cause the chair's immediate evasive action. As both sensors react alternatingly, the chair can get into an oscillatory motion, commonly known as "Braitenberg's oscillation" after [2]. In this specific situation, other frontally-mounted IR sensors take in "just go ahead" signals that invoke behaviors which can break the tie.

6.5   Priority Scheme

The priority arrangement is shown in the top right corner of Figure 6, where several output lines to the motors are joined by (s) nodes or suppression nodes. Input from the left of the node is suppressed and replaced by one coming in vertically whenever it is active. Inputs from the joystick take the highest priority in deciding which action the chair should take. The electronically and mechanically seamless interface between the joystick controller and the autonomy management system allows the chair to run as a standard power wheelchair simply by operating the joystick. The second highest priority is given to behaviors which take

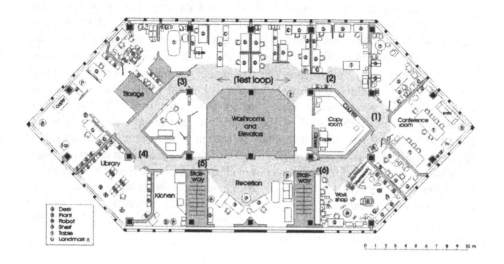

Fig. 8. The office space which contains the test loop.

in signals from left and right bumpers and some key frontal IR sensors. Behaviors are bundled up in Figure 6 with implied logical relationships among input lines to simplify the diagram. There are several groups of behaviors that mostly depend on signals from IR sensors for their invocation. These are followed by behaviors invoked by signals from the voice input system, followed by vision-driven behaviors as the lowest priority behavior groups. They are, in descending order of priority, depth map, vanishing point, and free area.

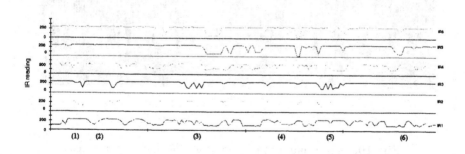

Fig. 9. Output of active infrared (IR) sensors

Figure 9 shows IR signals from a test run in which TAO-1 went around the test loop in our office floor shown in Figure 8 (shaded area). Note that signals from only 6 of the 12 IR sensors are plotted here. The x axis in Figures

10a through 10d shows the passage of time and its length corresponds to the time required to complete the loop from the workshop and back there counter-clockwise. Note that checkpoints (1) through (6) shown in Figure 8 are also marked on the diagrams. When there is no reflection from an obstacle, output of an IR is kept at 255. Depending on the strength of the reflected signal, a receptor may report lower values, 0 being the lowest. When the value becomes less than a threshold, the sensor would have "detected an obstacle." The threshold is set as a function of the speed of the chair, and in this specific test is set at 210, 180, and 150, for when the chair is running, at fast, medium, and slow speed, respectively. In another mode of obstacle detection using an IR, changes in value are monitored for several sensor cycles. If the change is sufficiently large, detection of an obstacle is reported. The IR sensors take in values at 64Hz and several consecutive values are compared. Once invoked, a behavior corresponding to a specific IR sensor generates a predetermined reactive motion, altering the speed and orientation of the chair.

6.6   Vision Processing

Inputs from 2 CCD cameras are alternatively processed through a single frame grabber into two primary vision planes of 256 x 128 pixels each at about 8 frame sets per second. Images in these primary vision buffers are averaged down to 64 x 32 pixel secondary vision plane by combining the left and right vision inputs after dividing each primary plane into left, center, and right. All vision processing described below occurs using image data in this secondary visual plane.

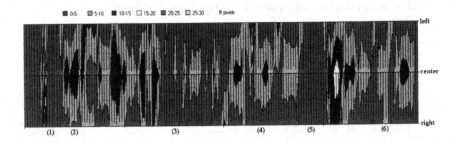

Fig. 10. Depth map parameters from the vision subprocess

Figure 9 plots three depth values (left, center, and right) in terms of the number of pixels in the secondary visual plane determined according to Horswill's habitat constraint vision processing [13]. In the absence of active bumper and IR invoked behaviors, the parameter set directly dictates the orientation and speed of the wheels.

Output from the vanishing point detector of the vision system is shown in Figure 9. The detector attempts to find a vanishing point in the secondary visual plane and outputs its x axis value when it finds one. The value 0 corresponds to the left-most angle in the visual plane, and 63 to the right-most. When it fails to come up with a vanishing point, value 99 is output. The combined horizontal viewing angle of the left and the right cameras is approximately 100 degrees.

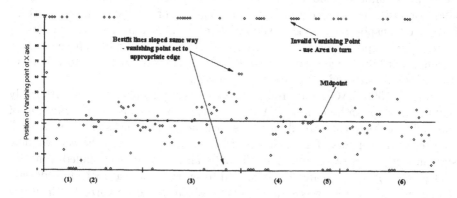

Fig. 11. Output of vanishing point detector.

Figure 12 depicts output from the area detector. The number of pixels representing free space in the left, center and right visual fields are calculated by the detector. Steering and speed of the chair are determined by the size of available space as in depth map processing. The behaviors associated with area detection are invoked only when all other behaviors are not invoked.

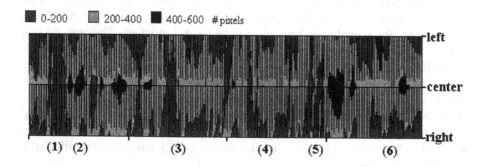

Fig. 12. Output of the area detector.

As the project proceeds the vision system will be enhanced to detect more objects and events such as outdoor landmarks, indoor landmarks that change in time, more complex and dynamic obstacles, and traffic signals in the path.

7   Lessons Learned so far from the Chair Project

Although the experience is still very limited, we can state that there is a strong expectation among the population for the development of an autonomous wheelchair for assisting and eventually fully taking care of the handicapped person's domestic transportation needs. We have demonstrated that the chair can travel at reasonable speeds through a standard North American office space with its peculiar attributes such as average width of passage ways, nature and volume of human traffic, size and orientation of doorways, etc.

In April 1996, TAO-1 was brought to a local shopping mall in Ottawa to freely roam around for an hour or so. TAO-1 skillfully skirted all internal structures of the mall such as escalators, flower pots, benches, signs, and showcases, as well as afternoon shoppers. TAO-1 and its rider visited stores as if he was window shopping or just strolling the mall. Virtually all fellow shoppers failed to notice that it was not driven manually. It tended to swerve downward when a sidewalk at the shopping center was slanted. This problem could be corrected in a few ways, and in fact, when we encountered the same problem with TAO-2 on a sidewalk in Japan, we successfully implemented one of the methods. This made us feel that with proper engineering to increase the chair's dependability, it can already serve as an autonomous chair for the severely handicapped in limited application areas, such as strolling or window shopping. Usability of the chairs in more constrained places such as smaller homes and limited office spaces would require further testing and revisions.

In the United States in the early 20th Century when automobiles began hitting humans on the street killing or injuring them, many cities and towns passed by-laws mandating each driver to have a "battler" running and waiving a flag (or a lantern after dark) in front of the car. This practically limited the maximum speed of automobiles to about 10 miles per hour. Of course, the practical application and enforcement of these bylaws met strong resistance from the reality, and the issue was replaced with other arguments or simply forgotten in many cases. Some of the by-laws are said to be still in effect. The episode tells a lot about human nature and what will likely happen to the fate of intelligent wheelchairs and similar "intelligent" machines that are meant to assist and help would-be human users in need. After the modest demonstration of TAO-2 in Japan, which was reported in local television news and several local newspapers, we have received inquiries for the chair's availability. Needless to say, it will be at least a few more years before even a modestly autonomous chair can be released for use by the handicapped at large and put into daily use only with affordable amount of support.

Maintenance would be another issue if we proceed, not to mention various public liability issues that, unfortunately but undoubtedly, will follow. The public liability issue is potentially a problem in introducing an autonomous or semi-autonomous wheelchair to the general public and this can become a hindrance to the effort to bring these technologies to the handicapped.

We are not at all optimistic about the efforts required to establish an infrastructure for physical and moral support that encompasses all these and other yet to be found issues. Nevertheless, we can foresee that we will be able to answer, in the near future, some of the sincere wishes that already come from people who would be most benefitted by the technology.

Getting into technical issues, the list of things yet to be done is still quite long. Landmark detection, for example, requires a lot more work. Although we have succeeded in navigating the chair to go through a series of landmarks arbitrarily chosen in the chair's present operational environment, this is still a far cry from being able to state that it can run freely in any environment traversable by a wheelchair by detecting landmarks.

Apart from these and other shortcomings, we feel the technology as it is, is already useful in real world applications by individuals with certain types of handicap. Persons with bodily contortions such as those who suffered polio in earlier life, or individuals with involuntary hand/arm movements such as patients of Parkinson's disease, now could travel through confined and narrow spaces such as corridors and doorways without assistance. Other interface mechanisms such as neck control and a voice recognizer would also make the introduction of the autonomous chair easier. Less handicapped users can use the chair as a manual power wheelchair whenever desired, while autonomy management can assist in mundane situations and emergencies.

Everybody with whom we have interfaced so far, from a passer-by at the shopping center where TAO-1 was tested, to fellow robotics researchers, several handicapped people and caregivers who heard about the project and came to see and even volunteered to try an early prototype, willing investors, and journalists all gave us positive feedback. They agree in principle that mobility should be provided as much as and as soon as possible to those who otherwise are not capable of going to places by themselves. Although the development is still far from complete, TAO-1 and 2 have so far been covered by several TV programs and a few dozen newspaper and magazine articles in Europe, Japan, USA, and Canada, indicating the keen level of interest the public has on this subject.

8   Conclusions

Two prototype autonomous wheelchairs based on commercially available motorized wheelchairs have been built using behavior-based AI. The initial prototyping went very rapidly and the size of the software is significantly smaller than control programs for similar vehicles operating in the real world environment implemented using conventional AI and robotics methodologies. One of the chairs is now capable of traveling to its indoor destinations using landmark-based navigation. The performance of the prototypes indicates there is a cautious possibility today to build a functional intelligent wheelchair that is practical and helpful to people with certain types and degrees of handicap.

9   Acknowledgements

Koichi Ide is responsible for the detailed design and most of the implementation of both TAO-1 and TAO-2. Reuben Martin and Richard Edwards assisted Ide in implementation and testing of both chairs.

References

1. D. A. Bell, S. P. Levine, Y. Koren, L. A. Jaros, and J. Borenstein. Design criteria for obstacle avoidance in a shared-control system. *RESNA '94*, Nashville, 1994.
2. V. Braitenberg. Vehicles: experiments in synthetic psychology. MIT Press, Cambridge, MA. 1984.
3. R. A. Brooks. A robust layered control system for a mobile robot. *IEEE Journal of Robotics and Automation*, RA-2, 1986.
4. R. A. Brooks. New approaches to robotics. *Science* 253, 1991, pp 1227–1232.
5. R.A. Brooks. Intelligence without reason. In *Proc. IJCAI-91*, Sydney, Australia, August 1991.
6. J. Connell and P. Viola. Cooperative control of a semi-autonomous mobile robot. *Robotics and Automation Conference*, 1990, pp 1118–1121.
7. J.D. Crisman and M.E. Cleary. Progress on the deictically controlled wheelchair. In Mittal et al. [15].
8. T. Gomi. Non-cartesian robotics. *Robotics and Autonomous Systems* 18, 1996, Elsevier, pp 169–184.
9. T. Gomi. Aspects of non-cartesian robotics. In *Proceedings of Artificial Life and Robotics (AROB'96)*, Japan, February 1996.
10. T. Gomi. Subsumption robots and the application of intelligent robots to the service industry. Applied AI Systems, Inc. Internal Report, 1992.
11. T. Gomi and K. Ide. The development of an intelligent wheelchair. In *Proceedings of IVS'96*, Tokyo, Japan, September 1996.
12. I. Harvey. Evolutionary robotics and SAGA: the case for hill crawling and tournament selection. In *Artificial Life 3 Proceedings*, C. Langton, ed., Santa Fe Institute Studies in the Sciences of Complexity, Proc. Vol XVI, 1992, pp 299–326.
13. I. Horswill. A simple, cheap, and robust visual navigation system. *Second International Conference on Simulation of Adaptive Behavior (SAB'92)*, Honolulu, Hawaii, 1992.
14. P. Maes. Behavior-based artificial intelligence. *Second International Conference on Simulation of Adaptive Behavior (SAB'92)*, Honolulu, Hawaii, 1992.
15. V. Mittal, H. Yanco, J. Aronis, and R. Simpson, eds. *Lecture Notes in AI: Assistive Technology and Artificial Intelligence*, 1998. This volume.
16. R. Pfeifer and C. Scheier. Introduction to "new artificial intelligence". Institut fur Informatik der Universitat Zurich, July 1996.
17. R.C. Simpson, S.P. Levine, D.A. Bell, L.A. Jaros, Y. Koren, and J. Borenstein. NavChair: an assistive wheelchair navigation system with automatic adaptation. In Mittal et al. [15].

18. L. Steels. Building agents out of autonomous behavior systems. *The Biology and Technology of Intelligent Autonomous Agents*, NATO Advanced Study Institute, Trento, 1993, Lecture Notes.

19. H. Wang, C-U. Kang, T. Ishimatsu, T. Ochiai. Auto navigation on the wheel chair. In *Proceedings of Artificial Life and Robotics (AROB'96)*, Beppu, Japan, February 1996.

20. H.A. Yanco. Wheelesley, a robotic wheelchair system: indoor navigation and user interface. In Mittal et al. [15].

21. H. A. Yanco, A. Hazel, A. Peacock, S. Smith and H. Wintermute. Initial report on Wheelesley: a robotic wheelchair system In *Proceedings of the Workshop on Developing AI Applications for the Disabled*, V.O. Mittal and J.A. Aronis, eds., held at the International Joint Conference on Artificial Intelligence, Montreal, Canada, August 1995.

Integrating Vision and Spatial Reasoning for Assistive Navigation*

William S. Gribble[1], Robert L. Browning[2], Micheal Hewett[2],
Emilio Remolina[2], and Benjamin J. Kuipers[2]

[1] Dept. of Electrical and Computer Engineering
The University of Texas at Austin Austin, TX 78712
U.S.A.
[2] Dept. of Computer Sciences
The University of Texas at Austin
Austin, TX 78712
U.S.A.
{grib,rlb,hewett,eremolin,kuipers}@cs.utexas.edu
url: http://www.cs.utexas.edu/users/robot

Abstract. This paper describes the goals and research directions of the University of Texas Artificial Intelligence Lab's Intelligent Wheelchair Project (IWP). The IWP is a work in progress. The authors are part of a collaborative effort to bring expertise from knowledge representation, control, planning, and machine vision to bear on this difficult and interesting problem domain. Our strategy uses knowledge about the semantic structure of space to focus processing power and sensing resources. The *semi-autonomous* assistive control of a wheelchair shares many subproblems with mobile robotics, including those of sensor interpretation, spatial knowledge representation, and real-time control. By enabling the wheelchair with *active vision* and other sensing modes, and by application of our theories of spatial knowledge representation and reasoning, we hope to provide substantial assistance to people with severe mobility impairments.

1 Introduction

The Intelligent Wheelchair Project is working to build an assistive agent capable of autonomous and semi-autonomous navigation in an initially unknown, dynamic world. We address this general goal by focusing on the specific problem of providing intelligent navigation assistance to profoundly mobility-impaired

* This work has taken place in the Qualitative Reasoning Group at the Artificial Intelligence Laboratory, The University of Texas at Austin. Research of the Qualitative Reasoning Group is supported in part by NSF grants IRI-9504138 and CDA 9617327, by NASA grant NAG 9-898, and by the Texas Advanced Research Program under grants no. 003658-242 and 003658-347.

V. O. Mittal et al. (Eds.): Assistive Technology and AI, LNAI 1458, pp. 179–193, 1998.
© Springer-Verlag Berlin Heidelberg 1998

people. People with moderate motor control can operate a standard powered wheelchair via a variety of mechanical and electronic interfaces. However, many people do not have enough motor control to operate such interfaces reliably or at all, and can be left dependent on others to make even the most basic movements from one place to another. Those with enough motor control to operate mechanical or joystick interfaces, but who are spastic or have some perceptual problems, may be unable to navigate in difficult environments without putting themselves in danger. The community of wheelchair users that will benefit most from our approach are those who do not have enough motor control to steer a wheelchair using traditional interfaces, or who can control a wheelchair to a limited extent but wish to have some assistance with difficult, tedious, or high-precision navigation tasks. The problem of assisting a user who has high cognitive function but severe motor and communication problems can be characterized as a case of taking noisy, error-prone, low-bandwidth control information from the human driver and providing smooth closed-loop control of the wheelchair's motion in service of specific navigation tasks.

1.1 Augmented Wheelchair Capabilities

We have built a mobile experimental platform with real-time vision from off-the-shelf computing hardware and a customized electric wheelchair. It has a stereo vision system with independent pan and tilt controls for each camera, an on-board computer, a laptop for user feedback and control, and a small embedded system to manage the wheelchair's sensor and motor hardware.

The wheelchair base is a TinMan II from the KISS Institute for Practical Robotics.[1] This is a Vector Velocity wheelchair, retrofitted with twelve infrared proximity sensors, seven sonars, and a small embedded computer which manages the drive systems and collects input from the sensors.

The principal on-board computer is a dual processor Pentium Pro machine running Debian Linux. Two frame grabber cards allow us to acquire images for dual-monocular or stereo image processing. The available computing power is sufficient to do substantial image processing on-board. User interaction (other than joystick commands) is handled through a laptop, also running Linux, which is connected to the main computer via an on board Ethernet network. Two CCD cameras provide our system's visual input, with each camera mounted on a Directed Perceptions pan-tilt head.

1.2 Tasks Addressed by This Work

The assistive tasks addressed by the Intelligent Wheelchair Project span a range of spatial scales and degrees of autonomy:

- **Hazard avoidance.** Obstacles, fast-moving people, and other hazards can cause problems for users who cannot react quickly to avoid them. An important task of the intelligent wheelchair is to detect and avoid such hazards using sonar, vision, and other available sensors.

---

[1] http://www.kipr.org

- **Semi-autonomous routines to assist in difficult maneuvers.** Many doorways, elevators, and wheelchair ramps have narrow entrances and tight quarters, requiring precise control of the wheelchair. A user who is spastic or has no control of his or her hands may find such precision navigation frustrating or impossible. For common situations such as the ones listed above, it may be appropriate to have *semi-autonomous routines*, triggered by the wheelchair driver. A semi-autonomous routine identifies correspondences between the local environment and a plan for handling the situation, and then follows the plan under executive supervision of the driver. Such routines depend on the wheelchair's driver to identify an appropriate context in which to execute them. A typical semi-autonomous routine might turn around a wheelchair and back it into an elevator under real-time visual control.
- **Providing "hands-free" control over medium distance travel.** Continuous manual control of a powered wheelchair can be taxing for people with limited motor control. Where a control strategy can be well-defined, such as along a sidewalk, down a hallway, or following a person, the driver could simply select from a set of available control routines. The wheelchair can use vision and other sensors to serve this type of simple autonomous control. The success of CMU's Navlab [13] and other automatic driving systems show that this is a feasible approach for environments with regular structure, especially when a human user is available to handle transitions and indicate when the environmental context is correct.
- **Navigating over large-scale distances using a spatial knowledge base.** Campus-scale environments can provide significant barriers to travel in a wheelchair. Automatic assistance in route-finding, combined with control strategies for medium-scale (following streets or sidewalks) and small-scale (traversing doors and wheelchair ramps) motion, can make it possible to semi-autonomously navigate a path from one point on campus to another.

Each of these problem domains is difficult in itself, and solving them all is a daunting challenge. However, assistive navigation has one important difference from the similar problem of autonomous mobile robot navigation which makes assistive navigation simpler in some respects and more difficult in others. That difference, of course, is that the wheelchair has a driver who is, at all times, in control (direct or executive) of the wheelchair's behavior.

1.3 The Wheelchair as Assistive Agent

We can conceive of the intelligent wheelchair as an *assistive agent* whose purpose is to serve tasks given to it by the driver. Having a human controller is a boon to the agent when it is confronted with environments or situations that are too complex for it to understand. Rather than fail gracelessly (as autonomous mobile robots do all too often) the agent can depend on manual control by the driver (if the driver has sufficient motor abilities) or engage in an interactive dialog to clarify the situation.

The driver also places requirements on the agent's behavior that makes some traditional mobile robotics techniques inappropriate. Many spatial-knowledge representation strategies rely on autonomous exploration of the environment to construct a map or model that is grounded in the agent's sensorimotor experience. For the most part, autonomous exploration is simply out of the question for an assistive agent. It must build its knowledge base through passive observation, or active strategies that do not interfere with the user's goals. Certain types of control strategies may also be inappropriate, particularly those that cause the wheelchair to make sudden changes in direction or rely on repetitive motion to build estimates of the location of particular features. Smooth motion and unobtrusive data collection are important for a user community whose differences already bring too much unwanted attention.

A prototypical assistive task of moderate complexity is the problem of helping to drive the wheelchair from one's office to fetch a cup of coffee on another floor. To assist a user having a near-complete lack of mobility, the agent must have enough knowledge to perform the task almost completely autonomously. In particular, it must be able to do the following things:

– **Represent knowledge about the world.** In order to plan effectively, the agent must have knowledge about the structure of large-scale space, such as the layout of buildings, campuses, and so on. Spatial knowledge can be *metrical, qualitative,* or some combination of the two. Other important knowledge is non-spatial, for instance, knowledge about the agent's state of being or the progress of its current activities, and must also be represented.
– **Receive and interpret user commands.** Persons with severe mobility impairments may not have enough motor control to reliably or completely specify their desires to the assistive agent. Interpreting noisy, imprecise command inputs requires knowledge of the agent's current situation and status, the user's likely action choices, and so on.
– **Build a plan from a library of primitive actions.** The user's command indicates a goal or desired state. The agent must use its knowledge of the world's state and its own abilities to construct a plan for achieving the specified goal. This plan will contain references to physical objects, landmarks, and other perceivable features of the environment embedded as direct and indirect objects of action. A significant part of the planning task is *attentional selection* of these objects (plan nouns).
– **Establish grounding connections between plan nouns and sensory input.** For each primitive action to be taken, the agent must find parts of the physical world which match the nouns referred to by that action. For example, a command to go through a door requires that a suitable door be located in the local area. All such relevant objects must be collected into representations of the local *visual space* and monitored with sensors to keep representations up-to-date.

These skills allow the intelligent wheelchair to accept a goal indicated by the user, construct a plan to achieve that goal, and connect the free nouns in the plan to physical objects in the world in order to perform the locomotor control needed to achieve the goal. In the remainder of this paper, we describe our approach to each subproblem.

2   Representing Knowledge About Large-Scale Space

Our approach for representing large-scale space is the Spatial Semantic Hierarchy (SSH) [8,9]. The SSH abstracts the structure of an agent's spatial knowledge in a way that is relatively independent of its sensorimotor apparatus and the environment within which it moves. In addition, the SSH integrates reactive behaviors, different spatial scales, and lends itself to a simple design for a human-wheelchair interface (see Section 3).

The Spatial Semantic Hierarchy (SSH) is an *ontological hierarchy* of representations for knowledge of large-scale space.[2] An ontological hierarchy allows multiple representations for the same kind of knowledge to coexist. Each level of the hierarchy has its own *ontology* (the set of objects and relations it uses for describing the world) and its own set of inference and problem-solving methods. The objects, relations, and assumptions required by each level are provided by those below it. The SSH is composed of 5 levels:

- **Sensorimotor.** The *sensorimotor level* of the agent provides continuous sensors and effectors, but not direct access to the global structure of the environment, or the wheelchair's position or orientation within it.
- **Control.** At the *control level* of the hierarchy, the ontology is an egocentric sensorimotor one, without knowledge of geographically fixed places in an external environment. A *distinctive state* is defined as the local maximum found by a hill-climbing control strategy, climbing the gradient of a selected feature, or *distinctiveness measure*. Trajectory-following control laws take the wheelchair from one distinctive state to the neighborhood of the next, where hill-climbing can find a local maximum, reducing position error and preventing its accumulation.
- **Causal.** The ontology at the SSH *causal level* consists of views, distinctive states, actions and schemas. A *view* is a description of the sensory input obtained at a locally distinctive state. An *action* denotes a sequence of one or more control laws which can be initiated at a locally distinctive state, and terminates after a hill climbing control law with the wheelchair at another distinctive state. A *schema* is a tuple $((V, dp), A, (V', dq))$ representing the (temporally extended) event in which the wheelchair takes a particular action A, starting with view V at the distinctive state dp, and terminating with view V' at distinctive state dq.[3]
- **Topological.** At the *topological level* of the hierarchy, the ontology consists of *places*, *paths* and *regions*, with connectivity and containment relations. Relations among the distinctive states and trajectories defined by the control level, and among their summaries as schemas at the causal level, are effectively described by the topological network. This network can be used

---

[2] In large-scale space the structure of the environment is revealed by integrating local observations over time, rather than being perceived from a single vantage point.

[3] Given a schema $S = ((V, dp), A, (V', dq))$, the *context*, *action* and *result* of S are V, A, and V', respectively. We intuitively read the schema S as, in context V, after executing A, expect V'.

to guide exploration of new environments and to solve new route-finding problems. Using the network representation, navigation among places is not dependent on the accuracy, or even the existence, of metrical knowledge of the environment.

– **Metrical.** At the *metrical level* of the hierarchy, the ontology for places, paths, and sensory features is extended to include metrical properties such as distance, direction, shape, etc. Geometrical features are extracted from sensory input, and represented as annotations on the places and paths of the topological network.

As an example, consider how the different SSH levels allow the wheelchair to represent spatial information about a university campus. The topological map associated with the campus is a graph where nodes (places) represent street intersections and edges represent the streets. This graph is useful for establishing a high level route from one place to another as well as to interact with the wheelchair's driver. Information from the control and causal levels is used to associate with each edge of the map the semi-autonomous routine needed for traveling a particular street. The local geometry of places in the map (i.e. street intersections) is represented by a detailed local metrical map (for example, an occupancy grid) which allows the wheelchair to find a plan for actually crossing a street.

3   Human Interface

The driver-wheelchair interface must allow the driver to communicate commands in a relatively natural language of spatial structure and action.[4] The wheelchair's spatial representation must facilitate the understanding of such commands and must present a sensible set of command options to the driver when required. For our purposes, the relevant properties of an interface device are the number of alternative commands the human driver may select from at any given time and the frequency with which selections must be made to guide the wheelchair. We abstract away the medium used to present the choices (for example, computer display, scanning menu, generated speech) and the device used to capture the choice (for example, joystick, pushbutton, "sip and puff" tube).

The different levels of the SSH lend themselves to a simple design for a useful interface:

– **Topological.** At the SSH topological level, the environment is represented as a graph (the topological map) of places and paths. Assuming the wheelchair has a sufficiently complete map, the driver can instruct the wheelchair where to go by inspecting the map, selecting a place, and saying, "Go there!". To display a topological map clearly, it is helpful to have a single frame of reference with a familiar orientation and a relative location for each place,

---

[4] These ideas developed through discussion with Sengul Vurgun and the other members of the Intelligent Robotics seminar, Fall 1997.

which can be used to map the graph onto the coordinates of the display medium.[5] Names alone could be used to specify destinations, but not all important destinations have concise names.

- **Causal.** At the SSH causal level, the environment is represented as a set of routes, each consisting of causal relations among context, action, and resulting situation. During the time while one action is executing, the wheelchair can display the possible actions at the next decision point (very likely a small set such as Turn Right, Turn Left, Go Straight, and Stop). Any time until the decision point is reached, the user can select one of these actions. Where a manual control law would require a continuous stream of (for example) joystick commands to adjust the wheelchair's course, the causal interface requires approximately two bits of information at each decision point.
- **Control.** At the SSH control level, the interface allows the driver to select the control law to follow from the current state. For example, the driver might wish to rotate clockwise to the next distinctive orientation or move along a corridor. The environment itself will provide the constraints that determine whether, for example, following a corridor requires a midline-following or wall-following control strategy. The SSH control level also includes a second interface: a simple interrupt signal that can be interpreted as "Stop!" or "Watch out!" or "Here it is!," depending on context. This signal is issued by the driver who may notice that a destination point has been reached, or a hazard is being approached, while the wheelchair may not notice it. Within the SSH framework, this signal can be thought of as signaling the end of a qualitatively uniform segment of the environment.

In this analysis, we have related the information-acquisition needs of the wheelchair user interface to the different representations for spatial knowledge in the SSH. In most cases, control of the wheelchair requires the selection of actions (control laws) from among small finite sets, without substantial time pressure. This reduces the communication bandwidth required between driver and wheelchair, when compared with direct control devices such as joysticks.

4   Constructing a Plan and Directing Attention

The intelligent wheelchair has limited processing power. High-bandwidth sensors, such as color cameras, provide enough information to make brute-force computing approaches impractical. Fortunately, human perception demonstrates that it is often unnecessary to process all of the data incident on sensors. Humans use *perceptual attention* to focus on a subset of the relevant stimuli, greatly reducing the amount of required sensory processing [14]. In order to similarly reduce the processing needs of the wheelchair, we use a perceptual attention strategy modeled on that of humans.

---

[5] A qualitative description of the SSH metrical information is used to create a visualization of the topological map.

Attention is directed to (i.e. processing power is allocated to) various portions of the perceptual input stream based on the current set of goals. For example, if the wheelchair is in the process of locating the elevator buttons, visual input will dominate the perceptual processing. Specific features (such as color, shape and orientation) and recognizable objects (such as hallways, doors, people and signs) receive preferential processing. Using these preferences, percepts can be filtered and prioritized before arriving at the cognitive level. This reduces the cognitive load by pre-identifying percepts that are the most interesting and relevant to the current goals of the system.

4.1   Plan Generation

Plans for movement in large-scale space can be constructed from the topological, causal, and control knowledge embodied in the SSH representation of space. Steps in the plan are the *actions* from each relevant schema at the SSH causal level, such as: *Move through the door, Turn left, Move down the hallway, Stop in front of the elevator.* Nouns in each step (*door, hallway, elevator*) indicate objects that are important to the task. The attention of the perceptual system can be focused mainly on these objects, allowing the plan to work as a top-down attentional selection mechanism. The system may use one or more of the available sensors to locate the important objects.

4.2   Coordinating the Perceptual Subsystems

We are designing a *sensor coordinator* to handle the problem of assigning perceptual tasks to multiple sensors. The process of allocating *sensor tasks* to sensors starts by collecting the objects mentioned in the current goals into the SACK (Set of Activated Concept Knowledge). Using a form of spreading activation [1], objects in the knowledge base that are related to the goal objects are added to the SACK. The level of activation needed to enter the SACK can be adjusted based on the current cognitive load of the system and the importance of the current goals. The contents of the SACK constitute the set of objects that are perceptually interesting to the system at the current time. The SACK is updated when the system's goals change.

Once the SACK has been updated, the sensor coordinator retrieves the *perceptual descriptions* (PD) of the objects in the SACK. Each PD is matched to sensor descriptions to find the appropriate set of sensors for perceiving that object. Sensor tasks are then allocated to each relevant sensor. Each task contains a sensor-specific description of the object.

To illustrate the interplay between cognition and perception, consider an intelligent wheelchair whose current goal is to enter an elevator. The elevator concept and its associated concepts elevator-door, elevator-controls and elevator-display will be members of the SACK. An elevator-control corresponds to an up or down button, whose perceptual description might be a "high-contrast circular region." The description is used by the vision system to locate the elevator button in the visual scene. At the same time, the description "obstacle within one foot" might be used by the sonar system to detect the wall near the elevator.

Current Research Problems in Sensor Coordination. The sensor coordinator described above is currently under development. Some of the questions we are addressing during this phase of the research include:

- **What is the perceptual description language?** The PD language is crucial to correct sensor allocation. The Sensor Coordinator uses it to match sensors and construct sensor tasks. It must be general enough to encompass all sensors, yet precise enough to encompass all objects.
- **Does the sensor coordinator interfere with the need for fine control of perceptual sensors?** Control laws for robot movement often require fine control of effectors and sensors, each of which can influence sensor data. How do we allow this while using the Sensor Coordinator, which adds a layer of abstraction to the perception process?
- **How do we prioritize sensor tasks?** There may be many sensor tasks allocated to each sensor. How do we prioritize them and how is the priority used?

5   Grounding Plan Nouns with Real-Time Vision

The SSH approach to control is built upon *trajectory following* control laws which cause the agent to follow a particular path defined by a feature or a set of features in the environment. Typical trajectory-following control laws might be "maintain a constant distance from the wall on your right," or "pass through the center of the door." A plan for moving through the environment consists of a sequence of such trajectory-following laws with *transition conditions* specifying when to switch from one to another. Once such a plan has been constructed, the first step in its execution is to connect the *plan nouns* needed by control laws (people, objects, features of the environment such as walls or doorways, etc.) to perceptions, in some sensor data stream, of corresponding objects in the physical world.

At the highest level, these nouns are placed into the SACK (see Section 4) and control and spatial-reasoning routines begin to reason about the position and properties of the corresponding physical objects. It is the responsibility of sensor subsystems to ensure that the correspondence between an activated concept and a physical object is made and maintained. The most demanding (and most powerful) sensor subsystem available to the wheelchair is the real-time vision system.

Our approach to vision has its lower levels based in the *active vision* paradigm [2,7] but significantly extends this model to handle a visual world that extends past a single camera's field of view. In this section, we discuss the ARGUS system [3,4] which provides the low-level foundations of our vision system, the *unified dynamic object memory* which addresses the *field of view problem*, and our ideas about integrating visual observations into coherent structural representations of *visual space* over time.

5.1   Focusing Visual Attention

A real-time, stereo, color image feed is more of a firehose of data than a manageable stream. Short of massive brute-force computation, processing all of the data incident on the sensors is impossible. Fortunately, the goals of a real-time vision system can be adequately served without processing all of the incoming data, or indeed processing much of it at all. By focusing processing on regions of the image likely to contain important objects, we can maintain grounding connections for objects in the SACK without being drowned by the firehose of visual data.

We identify two modes of attentive processing for the vision system: *top-down* or *task-directed attention*, which converts sensing goals into specific processing directives (for instance, "Find a fire extinguisher in the current field of view") and *bottom-up* or *pop-out attention*, which identifies and classifies unexpected, suddenly-appearing, or hazardous objects in the visual stream.

Our work to date has focused principally on *top-down* direction of attention, since the principal purpose of vision in our work is to ground symbols from the SACK which represent physical objects.[6] Restricting ourselves to tracking physical objects opens up a variety of problem constraints and optimizations.

Tracking primitive image features. The images of particular physical objects occupy restricted areas of the agent's field of view, meaning that they can be identified in a small subwindow of the whole frame. They move, scale, and deform continuously over time, meaning that they can be tracked by image-processing routines of moderate complexity (applied only to a small portion of the whole frame) coupled with well-understood tracking algorithms. Objects with resolution-invariant properties, such as color or straight edges, can be viewed and tracked through low-resolution subsampled images (which we call *attention buffers*) to reduce computation. When there are multiple objects of interest in the visual scene, they can be tracked independently through separate subwindows. Rather than applying image analysis routines to an entire video frame, a subwindow is extracted and analysis is performed in its restricted frame of reference. The inspiration for our development of window-based processing came from Pierce's "roving eye" experiments [12] but the same techniques have been successfully pursued elsewhere and are generally credited to Hager[6].

Tracking objects. Multiple *attention buffers* with different image processing algorithms can be used together to track an object with multiple features. We have developed a simple language to describe the features making up an object and their geometric relationships. We use a probabilistic technique to detect and correct tracking errors in individual features.

---

[6] Hazard avoidance and other tasks which detect pop-out phenomena can be recast as top-down tasks ("Find any object which is moving quickly") when necessary.

Stereoscopic tracking. Tracking features in a pair of cameras provides additional constraints which actually makes tracking easier. Constraints from epipolar geometry and from knowledge of the physical size of objects of interest can minimize correspondence problems and can help in detecting tracking errors (once an estimate of the 3-d size of an object has been made, that estimate can be used to make a sanity check on the results of image-space tracking). Since we are concerned with the qualitative spatial relations between objects rather than with precise reconstruction of the 3-d scene, camera calibration is not critical and dense depth-map generation is unnecessary. We use a simple pinhole projection model, a mechanically imprecise mounting rig, and point-feature triangulation to compute 3-d position estimates which are fine for our purposes.

5.2   The Field of View Problem

The primary goal of the wheelchair's vision system is to ground symbols representing relevant objects from the SACK. These objects may be in front of the wheelchair, behind it, to the side, or even above the wheelchair, and with several objects in the current SACK, perhaps several or all of these simultaneously. However, most sensors (and, in particular, vision sensors) can view only a limited portion of the environment at any instant.

Standard cameras can only see the full 360 degree spatial surround by physically shifting their field of view.[7] Typical lenses for machine vision applications have fields of view which range from 10 or less degrees horizontally to 120 degrees or more for "fisheye" lenses. Lenses with greater than approximately 65 degrees horizontal field of view suffer both from low spatial resolution (normal video capture hardware generally has on the order of 256K pixels per captured frame) and from image distortion that can be troublesome to remove in a real-time system. A camera with a narrow enough field of view to have low distortion and high resolution must direct its field of view intelligently if it is to have a chance of seeing all the relevant parts of the environment.

We want to use *closed-loop* control of the wheelchair to avoid the problems associated with out-of-date world models and unexpected changes in the local situation. Our experiments with *visual servoing* control of a hand-eye robot [5] and real-time vision have demonstrated that this is tractable on the available hardware [3,4]. However, with finite sensing hardware, we cannot guarantee that all relevant objects will be in view at any given time.

5.3   Unified Dynamic Object Memory

To circumvent the field of view problem for closed loop control, we use a *unified dynamic object memory system* to mediate between sensor inputs and control-signal calculations. The unified dynamic object memory integrates real-time visual perception of visible objects with short-term active memory traces of objects

---

[7] Some researchers have used special lenses or panoramic imaging techniques with mirrors, but the problem of focusing perceptual attention remains.

that are not currently in the field of view. When information about an object's position or motion is needed to compute a control signal, the unified object memory provides that information. The unified object memory connects directly to visual perceptions when objects are visible and to estimates otherwise. This is a direct generalization of the *visual servoing* approach, which uses direct observation of the image stream to control motion but cannot handle objects outside the field of view.

We use a Kalman filter (KF) estimate of object position and motion. The KF provides both a dynamic estimate of the object's state, which evolves even when no direct observations are being made, and a measure of the uncertainty in the state estimate, which grows over time if no direct observations are made. The agent's field of view is controlled by an *investigatory action scheduler* which examines position uncertainty estimates in the unified object memory and selects an object for direct observation. The scheduler tries to maximize a metric which takes into account both the reduction of uncertainty that would result from observing an object and the importance of that object to the current task. Selection of an appropriate attention policy for each significant object is a function of the current context, the nature of the task, and the recent behavior of the object in question.

Unified object memory and an independent investigatory action scheduler allow plans for locomotion to be carried out without regard to the limitations of the agent's field of view. Any objects or features of the environment that must be perceived to control the wheelchair will be represented in the persistent dynamic object memory. Control laws for the wheelchair can ignore the actions needed to locate or track objects as the chair moves, or to switch cameras between one object and the next. These actions are generic to any task needing to know the position or properties of an object, and can be handled by a single investigatory action scheduler for all tasks.

5.4   Representing Visual Landmarks over Larger Spatial Scales

The unified object memory determines and maintains object position estimates over short time scales and within local spaces. It does not address the issue of maintaining representations of these perceptions over longer time scales and greater distances. Sets of observations of the same regions from different vantages can be integrated into larger scale representations of an area, and these larger scale, longer term representations of the spatial layout of observed visual landmarks can support more sophisticated reasoning, including the planning of novel routes through untraveled regions, the determination of appropriate landmarks for piloting through these regions [10], the improved estimation self-location, and, as the representations become more descriptive with the assimilation of more information, the support of more critical evaluation of future input.

Building broader representations of visual landmarks. The positions of objects as measured by a sensor system are naturally encoded in coordinates relative to the observer (an *egocentric* frame of reference). While egocentric, metrical representations of landmark positions are useful, world-centered (*allocentric*) frames of reference are more appropriate for many applications, and *qualitative* information about position (such as left-right and front-back orderings) is often more robust than metrical information, and can be more reliably extracted from sensor data.

Allocentric representations can be constructed from collections of egocentric representations when those collections share enough landmarks to align their coordinate frames. An allocentric representation describes the positions of perceived objects within a local coordinate system from a "third person" perspective. Allocentric representations become useful when large spatial scales and observer motion make reasoning about accumulated egocentric data impractical. Allocentric maps can be related to one another in a loose hierarchy when inter-map relations are known; one related approach to this kind of arrangement can be found in the work of McDermott and Davis [11].

Planning with landmark information. Information about the position of a series of landmarks allows the agent to devise completely novel travel plans. If some chain of known visual landmarks covers regions between the current location and a goal location, the agent may be able to generate a successful plan using these landmarks as piloting beacons. This topic is discussed extensively in the context of outdoor navigation in [10].

Knowledge of the arrangement of local landmarks can also improve efficiency when planning on much smaller scales. For example, when approaching a known area, even if from a novel direction, recalled maps of a given environment can allow the agent to determine where it should expect to find any landmarks relevant to its next intended actions. This knowledge can substantially speed up visual search operations.

Determining the agent's location. By measuring its position with respect to some number of known visual landmarks, the agent can evaluate its current location in a global frame of reference. Depending on the nature of the cognitive representations of the area, this may result in an accurate quantitative estimate or a more general qualitative one. Even when the agent is hopelessly lost it can use its landmark knowledge combined with local sightings to generate a list of its most likely current locations. From this list, using knowledge of the associated areas, it can generate a plan of travels and observations which will identify the correct one.

Evaluating new input in context. When a new observation must be incorporated into existing models, determining the appropriate meaning of the observation is key. Consider the observation of a trash can in a substantially different location from that of another trash can recorded in one of the agent's cognitive maps of the same general area. What does this mean: "Did the trash can move?," "Is there another trash can?" or "Is the map in error?" If the latter, then "In what way?"

Coming up with a correct answer to these questions is a difficult task, but there are a number of ways to address the issue. To determine if the trash can moved, a visual query can be made asking if a trash can can be spotted in the old, predicted location. The accuracy of this sighting can even be double checked by issuing queries for other objects which the map indicates should be nearby. If another trash can is spotted, then it is quite likely that there are in fact two trash cans. If not, then the possibility that either the trash can moved, or that the map is in error can be evaluated by a sequence of visual queries, expanding outward from the expected observation angle of the trash can in order to determine the accuracy of the rest of the map. Furthermore, the possibility that the object in question just moved can be influenced by annotations in the agent's knowledge base indicating the relative expected permanence of various objects. For example, it might be recorded that the probability that a trash can has moved is much higher than that of a building, or an elevator. Furthermore, nodes in the egocentric and allocentric maps must record "support" information so that newer, more "vivid" observations (or lack thereof) that contradict the existing regional representations can override stale information.

6   Summary

For an agent to be a useful assistant in navigation, route finding, or other mobility-related tasks, it needs several things: an understanding of space, tools of representation and inference for reasoning about space and action, sensors to perceive objects and spatial relations, and effectors to move through the world. The Intelligent Wheelchair Project is working to bring together the relevant theory and practice from the various related fields of AI, vision, and robotics to build such an agent. In this paper, we have outlined our approach, which uses focused perceptual attention to allocate available sensors most effectively for vision-guided navigation.

As we complete assembly of the software and hardware portions of our experimental platform, we will incrementally bring our agent's capabilities on-line. We have demonstrated the basic technologies of real-time active visual control, path-following control of a mobile robot, and construction of a spatial semantic map from controlled exploration. Integration of these modules will provide significant challenges in the months ahead.

References

1. J. R. Anderson. Spreading activation. In J. R Anderson and S. M. Kosslyn, editors, *Tutorials in Learning and memory: Essays in honor of Gordon Bower.* W. H. Freeman, 1984.
2. D.H. Ballard. Animate vision. *Artificial Intelligence,* 48:57–86, 1991.
3. W.S. Gribble. Slow visual search in a fast-changing world. In *Proceedings of the 1995 IEEE Intl. Symposium on Computer Vision (ISCV-95),* 1995.
4. W.S. Gribble. ARGUS: a distributed environment for real-time vision. Master's thesis, University of Texas at Austin AI Lab, Austin, Texas, 1995.

5. W.S. Gribble, R.L. Browning, and B.J. Kuipers. Dynamic binding of visual percepts for robot control. Unpublished manuscript, 1997.
6. G.D. Hager. *Task-directed sensor fusion and planning: a computational approach.* Kluwer Academic Publishers, 1990.
7. A. Bandopadhay J. Aloimonos and I. Weiss. Active vision. *Int. J. Computer Vision*, 1:333–356, 1988.
8. B. Kuipers. A hierarchy of qualitative representations for space. In *Working papers of the Tenth International Workshop on Qualitative Reasoning about Physical Systems (QR-96)*. AAAI Press, 1996.
9. B. Kuipers and Y. T. Byun. A robot exploration and mapping strategy based on semantic hierarchy of spatial representations. *Journal of Robotics and Autonomous Systems*, 8:47–63, 1991.
10. B. J. Kuipers and T. Levitt. Navigation and mapping in large scale space. *AI Magazine*, 9(2):25–43, 1988. Reprinted in Advances in Spatial Reasoning, Volume 2, Su-shing Chen (Ed.), Norwood NJ: Ablex Publishing, 1990.
11. D. McDermott and E. Davis. Planning routes through uncertain territory. *Artificial Intelligence*, 22:107–156, 1984.
12. D.M. Pierce and B. Kuipers. Map learning with uninterpreted sensors and effectors. *Artificial Intelligence*, 92:169–227, 1997.
13. C.E. Thorpe, editor. *Vision and Navigation: the Carnegie-Mellon Navlab.* Kluwer Academic Publishers, 1996.
14. A. Treisman. Features and objects. In A. Baddeley and L. Weiskrantz, editors, *Attention: Selection, Awareness and Control.* Clarendon Press, 1993.

Speech and Gesture Mediated Intelligent Teleoperation

Zunaid Kazi, Shoupu Chen, Matthew Beitler, Daniel Chester, and
Richard Foulds

Applied Science and Engineering Laboratories,
University of Delaware/DuPont Hospital for Children,
1600 Rockland Rd., P.O. Box 269, Wilmington, DE 19803
U.S.A.
kazi@asel.udel.edu

Abstract. The Multimodal User Supervised Interface and Intelligent
Control (MUSIIC) project addresses the issue of telemanipulation of everyday objects in an unstructured environment. Telerobot control by
individuals with physical limitations pose a set of challenging problems
that need to be resolved. MUSIIC addresses these problems by integrating a speech and gesture driven human-machine interface with a knowledge driven planner and a 3-D vision system. The resultant system offers
the opportunity to study unstructured world telemanipulation by people
with physical disabilities and provides means for generalizing to effective
manipulation techniques for real-world unstructured tasks in domains
where direct physical control may be limited due to time delay, lack of
sensation, and coordination.

1 Introduction

A number of studies have shown the need for some general purpose manipulation aid for people with physical disabilities. Stanger *et al.* report in a review of
nine different task priority surveys that the top priority identified by users was
a device that could 'accommodate a wide range of object manipulation tasks in
a variety of unstructured environments' [27]. In an earlier quantitative study by
Batavia and Hammer, a panel of experts in mobility related disabilities ranked
fifteen criteria of importance in assistive technology [3]. The results for robotic
devices indicate that effectiveness and operability were the top two priorities.
In an informal study of user needs, a panel of people with disabilities strongly
supported the concept of a rehabilitation robot, but felt the existing interface
strategies were ineffective in offering the full potential of the device to a person
with a disability [10]. The panel strongly suggested that an effective rehabilitation robot system should:

- operate in an unstructured environment,
- require low mental load,

V. O. Mittal et al. (Eds.): Assistive Technology and AI, LNAI 1458, pp. 194–210, 1998.
© Springer-Verlag Berlin Heidelberg 1998

- provide maximum speed of operation,
- offer opportunities for use in many and varied environments (as opposed to a fixed workstation),
- be *natural to operate* (i.e. use functions that are easy and intuitive), and
- provide maximum use of the capabilities of the robot.

2   Background

While rehabilitation robotics research literature describes many demonstrations of the use of robotic devices by individuals with disabilities [12,9,13,2,4], many of the existing interface strategies, while important steps forward, have not met all of the desires of the user community. Traditional interfaces had generally taken two distinct approaches. One approach relies on the activation of pre-programmed tasks [26,12,11,30,14], while in a contrasting approach the user is directly manipulating the robot much like a prosthesis [33,18]. The command based system is limited by the need for a reasonably structured environment, and the fact that the user needs to remember a large number of often esoteric commands. On the other hand, direct prosthetic-like control allows operation in an unstructured operation, but tends to impose significant cognitive and physical loads on the user. In the case of a user with some physical disability, her range of motion may even preclude her from using such a device.

These limitations as well as the studies on user requirements prompted the investigation into the development of a reactive, intelligent, instructible teler-obot controlled by means of a hybrid interface strategy where the user is part of the planning and control loop. This novel method of interface to a rehabili-tation robot is necessary because in a dynamic and unstructured environment, tasks that need to be performed are sufficiently non-repetitive and unpredictable, making human intervention necessary.

However, the design of the instructible aspect of the assistive robot system requires careful design; simple command based interfaces may be inadequate. The limitations of a command-based interface were discussed by Michalowski *et al.* [21]. They propose greatly expanding the capability of the robot to not only recognize spoken words, but also understand spoken English sentences.

In a continuation of this work, Crangle *et al.* provided an example where the user spoke the sentence, 'Move the red book from the table to the shelf' [6,7]. The proposed system would recognize the spoken sentence and understand the meaning of the sentence. The system would have a knowledge of the immediate world so that the robot knew the locations of the table and shelf, as well as the placement of the book on the table. While the use of such natural language interfaces is extremely interesting, and would offer great benefit, the limitations are many. The requirement that the world be entirely structured so that the robot knows precisely where every item is, is likely to be too demanding, and there are many unsolved issues in natural language processing. In addition, the inclusion of a vision system to accommodate a less structured environment will require the ability to perform object recognition.

A different approach to command-based robot operation was proposed by Harwin *et al.* [16]. A vision system viewed the robot's workspace and was programmed to recognize bar codes that were printed on each object. By reading the bar-codes and calculating the size and orientation of the bar-code, the robot knew the location and orientation of every item. This was successful within a limited and structured environment. This system did not easily lend itself to a variety of locations and was not able to accommodate the needs of individuals with disabilities in unstructured environments. It did, however, demonstrate the dramatic reduction in machine intelligence that came by eliminating the need for the robot to perform object recognition.

3   Objective and Illustration

The previous discussions raised the following issues that needed to be solved for a practical real-world assistive robot, as well for any manipulation environment where the physical control of the potential user is less than optimal, and the environment involves known tasks and objects which are used in an inherently unstructured manner.

- The robot must be operable in an unstructured environment,
- There must be a natural and flexible Human-Computer interface, and
- The system must have a degree of autonomy to relieve user load.

These issues led to the development of MUSIIC (Multimodal User Supervised Interface and Intelligent Control), an intelligent, instructible assistive robot that uses a novel multimodal (speech and gesture) human-machine interface built on top of a reactive and intelligent knowledge-driven planner that allows people with disabilities to perform a variety of novel manipulatory tasks on everyday objects in an unstructured environment.

MUSIIC also shows that by combining current state of the art in natural language processing, robotics, computer   vision, planning, machine learning, and human-computer interaction, building a practical telemanipulative robot without having to solve the major problems in each of these fields is possible. Difficulties involving full text understanding, autonomous robot-arm control, real-time object recognition in an unconstrained environment, planning for all contingencies and levels of problem difficulty, speedy supervised and unsupervised learning, and intelligent human-computer interfaces, illustrate some of the open issues. Current solutions to each of these problems, when combined with each other and with the intelligence of the user, can compensate for the inadequacies that each solution has individually. We claim that the symbiosis of the high level cognitive abilities of the human, such as object recognition, high level planning, and event driven reactivity, with the native skills of a robot can result in a human-robot system that will function better than both traditional robotic assistive systems and current autonomous systems.

MUSIIC is a system that can exploit the low-level machine perceptual and motor skills and excellent AI planning tools that are currently achievable, while allowing the user to concentrate on handling the problems that they are best suited for, namely high-level problem solving, object recognition, error handling and error recovery. By doing so, the cognitive load on the user is decreased, the system becomes more flexible and less fatiguing, and is ultimately a more effective assistant.

A very simple illustration (Figure 1 and Figure 2) describes how our proposed system functions in a real-world scenario. The user approaches a table on which there are a straw and a cup The user points to the straw, and says, *that's a straw*. The user points to the cup and says *Insert the straw into that*, indicating that the straw must be insterted in the object indicated by 'that.' The robot arm then picks up the straw and inserts it into the cup.

4   MUSIIC Architecture

The previous sections lead naturally to a description of the essential components of the MUSIIC system. We require a **Planning Subsystem** that will interpret and satisfy user intentions. The planner is built upon object oriented knowledge bases that allow the users to manipulate objects that are either known or unknown to the system. A **Human-Computer Interface Subsystem** parses the user's speech and gesture to invoke task planning by the planner. An active stereo-**Vision Subsystem** is necessary to provide a snap-shot of the domain; it returns object shapes, poses and location information without performing any object recognition. The vision system is also used to identify the focus of the user's deictic gesture, currently implemented by a laser light pointer, returning information about either an object or a location. The planner extracts user intentions from the combined speech and gesture multimodal input. It then develops a plan for execution on the world model built up from the a priori information contained in the knowledge bases, the real-time information obtained from the vision system, the sensory information obtained from the robot arm, as well as information previously extracted from the user dialog.

4.1   Planning Subsystem

The planning subsystem is built on an object-oriented knowledge base called the **WorldBase** which models knowledge of objects in a four-tier abstraction hierarchy. In addition, there is a knowledge-base of domain objects called the **DomainBase** which models the actual work-space by incorporating information obtained from the vision subsystem and user input. There is also a user extendible library of plans (PlanBase) ranging from primitive robot handling tasks to more complex actions and purposeful actions. An intelligent and adaptive planner uses these knowledge bases and the plan library in synthesizing and executing robot plans.

Object Inheritance Hierarchy Objects are represented in an increasingly specialized sequence of object classes in an inheritance hierarchy. We have developed a four tiered hierarchy, where object classes become increasingly specialized from the top level class to the bottom level class as shown in Figure 3. The four classes are:

- **Generic Level** - This is the top level tier in the hierarchy of objects. This class is invoked during the planning process when nothing is known a priori about the object except the information obtained from the vision system, which includes object location, orientation, height, thickness, width and color.
- **Shape-Based Level** - This class of objects is the second tier in the class hierarchy. Since the domain of this research is robotic manipulation, the primary low level robot action of grasping is directly affected by the shape of the object. Current classes available at this level are: Generic, Cylindrical, Conical, Cuboid.
- **Abstract-Object Level** - The third tier constitutes general representation of commonly used everyday objects, such as a cup, can or box. These objects are derived from the Shape-Based classes. Object attributes such as approach points and grasp points are further specialized at this level. Constraints are added at this level to restrict the mode of manipulation.
- **Terminal Level** - This is the fourth and final tier of the abstraction hierarchy. These objects are derived from the Abstract-Object classes and refers to specific objects which can be measured uniquely such as the user's own cup, a particular book, or the user's favorite pen. Information from the vision system as well as user input may be used to instantiate attributes for these objects. All attributes are fully specified, and this allows a more accurate execution of user intentions by the robot.

Fig. 1. That's a straw

Fig. 2. Insert the straw into that

Object Representation Each object has a set of attributes that assists the planner in developing correct plans. The instantiations of these attributes are dependent upon to which abstraction hierarchy does one object belong. Table 1 describes these attributes.

Planning and Plan Representation The plan knowledge base, PlanBase, is a collection of STRIPS-like plans [24,25] and the planner is based on a modified STRIPS-like planning mechanism. The main difference between conventional STRIPS-like planning and our system is that we take full advantage of the underlying object-oriented representation of the domain objects, which drives the planning mechanism. Plans in this model are considered to be general templates of actions, where plan parameters are instantiated from both the WorldBase and the DomainBase during the planning process.

Plans in the plan library have the following general format:

- **Name**: This slot defines the name of the plan as well as the parameters with which this plan is invoked. Given the manipulative domain of this planner, these parameters are either an object name or a location name or both. These parameters are global variables, which are initialized by the planner at the beginning of the plan synthesis process, and the initialized values are used by the plan and any subactions of the plan.
- **Plan Type**: This slot identifies the type of the plan. Plans can be of three types: primitive, complex, and user defined. Primitive and complex plans are system encoded plans that are defined a priori. Primitive plans are those that are executed directly as a robot operation (i.e., their plan body contains only one subaction). Complex plans are plans whose plan body contains a list of plans, both primitive and complex.

Table 1. System Attributes.

Attribute	Contents	Comments
Object Name	Name of the object	For example: cup
Object Class Name	Name of the class to which the object belongs	For example: Cylindrical
Object Shape	Pointer to a data structure representing shape	
Object Color	floating-point value representing hue	
Height	floating-point value	
Width	floating-point value	
Thickness	floating-point value	
Location	Set of twelve floating-point values representing the Cartesian coordinates for orientation and position.	
Grasp Point	Set of twelve floating-point values representing the Cartesian coordinates for grasp position.	
Approach Point	Set of twelve floating-point values representing the Cartesian coordinates for approach position.	
Default Orientation	Set of twelve floating-point values representing the Cartesian coordinates for default orientation.	
Plan Fragments	Pointer to a list of symbolic expressions	Plan fragments are incorporated into plans formed by the planner. Certain tasks may be specific to an object, and plan fragments for those tasks may be associated with the object in question in order to facilitate correct planning.
Attribute	Pointer to a list of attributes	For any additional attributes that the user may specify, such as size, weight and malleability.

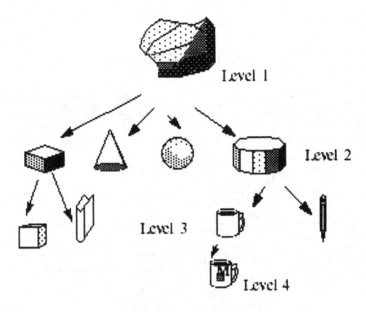

Fig. 3. Object Inheritance Hierarchy

- **Preconditions**: The preconditions represent what must be true before the plan can be invoked. During the interpretation of user instructions, the planner makes no attempt to make a precondition true. This restriction is lifted during the replanning and plan adaptation process.
- **Plan Body**: The plan body contains the subactions that are executed to make true the goals of the plan. This structure allows the MUSIIC planner to apply hierarchical planning.
- **Goals**: The goals specify what this action is intended to achieve.

Planning and Execution Our multimodal human-machine interface and object-oriented representation allows the user to interact with the planning system at any level of the planning hierarchy, from low-level motion and grasp planning to high-level planning of complex tasks. Synthesized plans are supplemented by user intervention whenever incomplete information or uncertain situations prevent the development of correct plans. The user does this by taking over control of the planning process or by providing necessary information to the knowledge bases to facilitate the development of a plan capable of handling a new or uncertain situation. Furthermore, incomplete sensory information may be supplemented by user input, enabling the planner to develop plans from its plan library without the need for extensive user intervention during plan execution.

MUSIIC can adapt previously learned plans and tasks to new situations and objects. Previously synthesized or learned plans can be modified to operate on new tasks and objects. Furthermore, MUSIIC can also react. When things go wrong, the system tries to determine the cause of error and autonomously replans to rectify the problem in simple cases such as when object placement is inaccurate. If, however, it is not able to ascertain the cause of failure, the system then engages in a dialogue with the user who takes over the plan correction and replanning procedures.

The MUSIIC planner also has the ability to learn from user instruction. The learning mechanism for MUSIIC is supervised learning. In this process, the user herself is in charge of determining what the system learns. This learning process can be either off-line or on-line. On-line learning is when the user is actually moving the manipulator through a set of tasks. In the case of off-line learning, the user simply enumerates a list of actions that make up the plan body of the plan that is to be learned.

4.2   Human-Computer Interface Subsystem

The HCI subsystem employs a multimodal input schema where users of our system point to indicate locations and speak commands to identify objects and specific actions. The combination of spoken language with pointing performs a critical disambiguation function. It binds the spoken words in terms of nouns and actions to a locus in the physical work space. The spoken input supplants the need for a general purpose object-recognition module in the system. Instead, 3-D shape information is augmented by the user's spoken word, which may also invoke the appropriate inheritance of object properties using MUSIIC's hierarchical object-oriented representation scheme.

Restricting Speech and Gesture In order to devise a practical command-input interpretation mechanism, we restricted both the nature of our speech input and our gesture input. While a fully fledged natural language system combined with a state-of-the-art gesture-recognition mechanism may allow the user more expressive power, developments in these two areas make this a distant goal. At the same time, the requirements of the domain places some constraints on the choice of modalities and the degree of freedom in expressing user intentions. A more practical alternative for use as an assistive telerobotic device is a multimodal combination of speech and pointing, where the input speech is a restrictive subset of natural language, which we may call a pseudo-natural language (PNL). We can then apply model-based procedural semantics [6], where words are interpreted as procedures that operate on the model of the robot's physical environment. One of the major questions in procedural semantics has been the choice of candidate procedures. Without any constraints, no procedural account will be preferred over another and there will not be any shortage of candidate procedures. The restrictive PNL and the finite set of manipulatable objects in the robot's domain provide this much-needed set of constraints. Similarly, the needs of users with disabilities also restrict the choice of gestures. Our gesture of choice is deictic gesture, which is simply pointing. In the general case, not only does pointing have the obvious function of indicating objects and events in the real world, it also plays a role in focusing on events/objects/actions that may not be objectively present [20]. The choice of pointing allows us to use any number of devices, not restricted to the hand, to identify the user's focus. While our test-bed uses a laser pointer to identify the user's focus of intentions, any device that is able to indicate a domain object, such as an eye tracking system or mouse on a control panel, can be used.

Natural Language Processing in MUSIIC We have adopted a restricted form of phrase-structure grammar for MUSIIC. While it does not have the full expressive power of natural language, the grammar of MUSIIC allows the user to operate the robot using utterances that are like natural language. The general syntax of MUSIIC is of the form VSO, where V stands for a verb, S stands for the subject, and O stands for the object. Given that the domain of MUSIIC is manipulatory, the subject is usually a domain object that is going to be manipulated, and the object of the utterance usually refers to a specific location that is the target of the manipulatory task. Assertions are also encoded in the grammar. This restricted format for the grammar allows us to use model-based procedural semantics to unambiguously chose a desired procedure for execution by the robot arm. Details of the grammar can be found in [17].

The parsed string is then sent to the supervisor whose control language then interprets the user instructions as robot control commands. The basic language units are calls to procedures that are either primitives made accessible from the underlying C++ modules, or calls to other procedures.

The supervisor's command language has a large number of these primitives, which can be classified into two general types: external and internal. External primitives are those that can be invoked by the user through a user instruction, while internal primitives are used by the supervisor for internal operations. A full list of the primitives in the command language is provided in [17]. Like all other data structures of the language, primitives are encoded as C++ classes with customized evaluation functions. During the interpretation process, when the interpreter comes across a syntactic unit, it looks it up in the symbol table, returns its value, and then evaluates the returned value. For example, grasp is a primitive of the external type. When the interpreter finds the word 'grasp' in an expression, it looks it up in the symbol table, where the value of 'grasp' is the primitive grasp. This primitive is then evaluated which ultimately generates a sequence of control commands that are sent by the supervisor to the robot arm.

4.3 Vision Subsystem

The vision subsystem allows MUSIIC to determine the three-dimensional shape, pose and location of objects in the domain. No object recognition is performed. The vision requirement is to provide the knowledge-based planning system with the parameterized shape, pose, orientation, location and color information of the objects in the immediate environment. This information can then be used to fill slots in the object-oriented representation and to support planning activities. The vision system plays a role in the user-interface as well. A pointing gesture, using a laser light pointer allows a user of MUSIIC to indicate her focus of interest. The vision subsystem identifies the spot of laser light and processes information relating to the object of choice.

The MUSIIC vision system utilizes feature-based stereo vision in order to recover 3-D world from 2-D images. Color images are used for the process of extracting features, which in our case are 2-D edges. After having extracted the features, a chain-code matching algorithm is used to extract the 3-D information about the world environment from them. The recovered 3-D information is further processed through Hough transformation to provide the planner with the orientations and positions of the objects of interest.

Color Image Processing Color has been increasingly used in machine vision in recent years for object discrimination [1,29,31]. In task planning for an assistive robot, color of the working environment is one of the key elements to build an efficient knowledge base system.

Images captured in our system are first decomposed in R, G, and B space. Thresholding is performed on the resultant R, G, and B images to separate the objects from the background. The threshold is determined by using the intensity histogram computed in each of the three color domains and results in binary images. These binary images are further processed to extract edges, which are then used to recover 3-D information.

While the R, G, and B components are employed to separate objects from the background, it is often difficult to identify a *color* by looking at these values directly. Instead, a transformation from the RGB space to a non-RGB psychological space has been shown to be effective in color perception. This non-RGB space is composed of 3 components, hue (H), saturation (S), and intensity (I). Instead of using the three values of R, G, and B, the single value of hue (H) can be used to label objects in this transformed space. While there exists many methods of transformation, this paper adopts a scheme described in [15].

Stereo Vision The main object of the stereo vision system is to provide the task planner with spatial information about objects in the domain, which may include size, shape, location, pose, and color. Features extracted from two 2-D images are subjected to a matching algorithm to find corresponding edge pairs. Existing algorithms may be classified into two major categories: feature-based (semantic features with specific spatial geometry) and area-based (intensity level as the feature) [8,32,28,23,22]. While a feature-based matching algorithm is fast and reliable, its processing is more complicated. We have developed a simple match algorithm that first transforms a 2-D edge signal to an 1-D curve through chain codes representation and then reduces the match dimension by one.

In 2-D images edges are planar curves and consequently edge curve matching is a 2-D operation. Li and Schenk proposed a $\Psi - S$ transformation such that a 2-D curve can be represented as a 1-D curve [19]. This $\Psi - S$ transformation, which is basically a continuous version of the chain code representation, transforms the

2-D edge-matching matching problem into a 1-D matching problem. We have used a similar chain-code technique to transform 2-D edges to 1-D curves.

The chain codes from the two images are fed into a *bipartite match network* [5]. An iteration process is performed to find the matching pairs in the two images. Once the matching process has been completed, calibrated camera transformation matrices can be used to recover the 3-D information of the images. A Hough transformation of this 3-D information then gives the spatial attributes of objects, such as height and width.

5   Implementation and Results

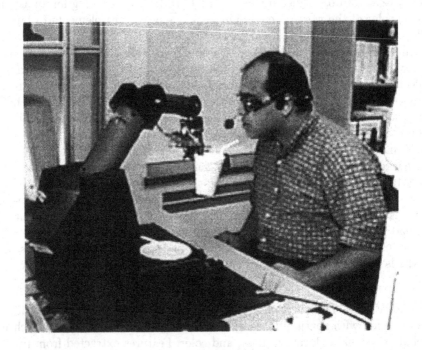

Fig. 4. System Set-Up

As shown in Figure 4, the actual hardware setup includes a vision subsystem, containing a pair of color cameras, an SGI workstation, and associated vision software, a six degree of freedom Zebra ZERO robot and controller, a speech recognition subsystem, and the planning and knowledge base system. These reside in different computing platforms and communicate with each other through Remote Procedure Calls (RPC) components. The operation of MUSIIC is illustrated through two annotated task scenarios. These scenarios involves the task of inserting a straw into a cup and bringing the cup to the user. The workspace contains a cup and a straw and the World Base contains entries for 'straws' and 'cups'.

5.1   Scenario One

Instruct the system to load in the plan library.

User: "Load plans"

MUSIIC: "Plan loading complete"

Instruct the system to synchronize the various system components.

User: "Synchronize"

MUSIIC: "Synchronization complete"

User: "Home"

The robot then moves to its home configuration.

MUSIIC: "Home successful"

User: "Scan"

The vision system generates object size, position, orientation and color information.

MUSIIC: "Scanning complete"

Instruct the vision system to transfer the information to the planning subsystem to build up the DomainBase.

User: "Load domain"

MUSIIC: "Domain loading complete"

The user points to the straw while simultaneously saying the word "straw".

User: "That's a straw"

MUSIIC: "Looking for the straw"

MUSIIC searches the WorldBase for the "straw".

MUSIIC: "I found the straw"

User points to the cup and identifies it to the system.

User: "That's a cup"

MUSIIC: "Looking for the cup"

MUSIIC: "I found the cup"

Instruct the robot to insert the straw into the cup.

User: "Insert the straw into the cup"

MUSIIC inserts the straw into the cup. On success:

MUSIIC: "I am ready"

Instruct the robot to bring the cup to the user.

User: "Bring the cup"

The arm approaches the cup and grasps it by the rim. It then brings the cup to a predetermined position that is accessible to the user (Figure 5).

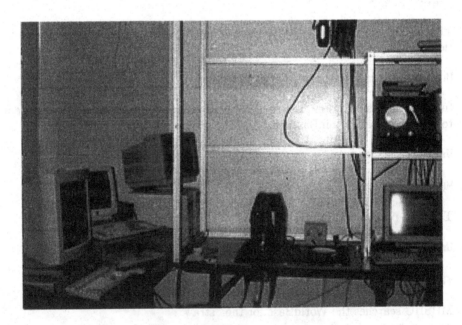

Fig. 5. After 'bring the cup'

5.2   Scenario Two

User: "Bring that"

Here we present an alternate scenario where the user did not explicitly identify the cup. The user points to the cup while simultaneously saying the verbal deictic 'that.' The vision system continuously records any identified spot along with time-stamp values that mark the time when the spot was recorded. The speech system also time-stamps utterances and these values are used to determine the location of the spot that was generated when a verbal deictic such as 'that' is spoken. The system then finds the object that is in that specific location and invokes the 'bring' task. Since the object has not been identified specifically, planning is based on general principles. Instead of grasping it by the rim the arm grasps the cup along its width and brings it to the user.

6   Conclusions

Human intervention as well as an intelligent planning mechanism are essential features of a practical assistive robotic system. We believe our multimodal robot interface is not only an intuitive interface for interaction with a three-dimensional unstructured world, but it also allows the human-machine synergy that is necessary for practical manipulation in a real world environment. Our novel approach of gesture- speech based human-machine interfacing enables our system to make realistic plans in a domain where we have to deal with uncertainty and incomplete information.

7   Acknowledgments

Work on this project is supported by the Rehabilitation Engineering Research Center on Rehabilitation Robotics, National Institute on Disabilities and Rehabilitation Research Grant #H133E30013, Rehabilitation Services Administration Grant #H129E20006 and Nemours Research Programs.

References

1. A. Andreadis and P. Tsalides. Coloured object recognition using invariant spectral features. *Intelligent and Robotic Systems*, 13:93–106, 1995.
2. D. C. Bacon, T. Rahman, and W. S. Harwin, editors. *Fourth International Conference on Rehabilation Robotics*, AI DuPont Institute, Wilmington, Delaware, USA, 1994. Applied Science and Engineering Laboratories.
3. A. Batavia and G. Hammer. Consumer criteria for evaluating assistive devices: Implications for technology transfer. In *12th annual Conference on Rehabilitation Technology*, pages 194–195, Washington DC, 1989. RESNA Press.
4. Bath Institute of Medical Engineering. *Fifth International Conference on Rehabilation Robotics*, Bath University, Bath, UK, 1997.

5. S. Chai and W Tsai. Line segment matching for 3d /computer vision using a new iteration scheme. *Machine Vision and Applications*, 6:191–205, 1993.
6. C. Crangle, L. Liang, P. Suppes, and M. Barlow. Using English to instruct a robotic aid: An experiment in an office-like environment. In *Proc. of the International Conference of the Association for the Advancement of Rehabilitation Technology (ICAART -88)*, pages 466–467, Montreal, Canada, 1988.
7. C. Crangle and P. Suppes. *Language and Learning for Robots*. CSLI Publications, Stanford, CA, 1994.
8. J. Cruz, G. Pajares, J. Aranda, and J. Vindel. Stereo matching technique based on the perceptron criterion function. *Pattern Recognition Letters*, 16(9):933–944, 1995.
9. R. A. Foulds, editor. *Interactive Robotics Aids-One Option for Independent Living: an International Perspective*. Monograph 37. World Rehabilitation Fund, 1986.
10. R. A. Foulds. Multimodal user direction of a rehabilitation robot. In *Rehabilitation Research Center on Robotics to Enhance the Functioning of Individuals with Disablities*, volume 2, pages 41–61. Center for Applied Science and Engineering, University of Delaware, 1993.
11. C. Fu. An independent vocational workstation for a quadriplegic. In R. Foulds, editor, *Interactive Robotics Aids-One Option for Independent Living: an International Perspective*, Monograph 37, page 42. World Rehabilitation Fund, 1986.
12. M. Gilbert and R. Foulds. Robotics at the Tufts New England Medical Center. In *10th Annual Conference on Rehabilitation Engineering*, pages 778–780, San Jose, California, 1987. RESNA Press.
13. M. Gilbert and J Trefsager, editors. *Proceedings of the 1990 International Conference on Rehabilitation Robotics*, Wilmington, Delaware, 1990. Applied Science and Engineering Laboratories.
14. J. Hammel, M. Van der Loos, and I. Perkash. Evaluation of DeVAR-IV with a quadriplegic employee. Technical report, Rehabilitation Research and Development Center, Palo Alto, California, 1991.
15. R. Haralich. *Computer and Robot Vision*. Addison-Wesley, 1993.
16. W. Harwin, A. Ginige, and R. Jackson. A potential application in early education and a possible role for a vision system in a workstation based robotic aid for physically disabled persons. In R Foulds, editor, *Interactive Robotic Aids-One Option for Independent Living: An International Perspective*, Monograph 37, pages 18–23. World Rehabilitation Fund, 1986.
17. Z. Kazi. *Integrating Human-Computer Interaction with Planning for a Telerobotic System*. PhD thesis, University of Delaware, Department of Computer and Information Sciences, 1997.
18. H. Kwee, M. Thonninsen, G. Cremers, J. Duimel, and R. Westgeest. Configuring the Manus system. In *RESNA International Conference, 1992*, pages 584–587, Toronto, Canada, 1992. RESNA Press.
19. J. Li and T. Schenk. Stereo image matching with sub-pixel accuracy. Technical paper, ACSM ASPRS, 1991.
20. D. McNeill. *Hand and Mind : What Gestures Reveal about Thought*. The University of Chicago Press, 1982.
21. S. Michalowski, C. Crangle, and L. Liang. Experimental study of a natural language interface to an instructable robotic aid for the severely disabled. In *10th Annual Conference on Rehabilitation Technology*, pages 466–467, Washington, DC, 1987. RESNA Press.
22. M. Okutomi and T. Kanade. A stereo matching algorithm: an adaptive window based statistical model. *Systems and Computers in Japan*, 23(8):26–35, 1992.

23. B. Ross. A practical stereo vision system. In *IEEE Computer Vision and Pattern Recognition*, pages 148–153, 1993.

24. E. D. Sacerdoti. The non-linear nature of plans. In *International Joint Conference on Artificial Intelligence*, pages 1123–1145, Tbilisi, USSR, 1975.

25. E. D. Sacerdoti. *A Structure for Plans and Behaviour*. American Elsevier, New York, 1977.

26. W. Seamone and G. Schmeisser. Evaluation of the APL/JHU robotic arm workstation. In R. Foulds, editor, *Interactive Robotics Aids-One Option for Independent Living: an International Perspective*, Monograph 37, page 51. World Rehabilitation Fund, 1986.

27. C. A. Stanger, C. Anglin, W. S. Harwin, and D. P. Romilly. Devices for assisting manipulation: A summary of user task priorities. *IEEE Transactions on Rehabilitation Engineering*, 2(4):256–265, 1994.

28. C. Stewart and J. MacCrone. Experimental analysis of a number of stereo matching components using lma. In *International Conference on Pattern Recognition*, volume 1, pages 254–258, 1990.

29. S. Tominaga. Colour classification method for color images using uniform colour space. In *IEEE International Conference on Pattern Recognition*, pages 803–807, 1990.

30. M. Van der Loos, J. Hammel, D. Lees, D. Chang, I. Perkash, and L. Leifer. Voice controlled robot systems as a quadriplegic programmer's assistant. In *13th Annual RESNA Conference*, pages 327–328, Washington DC, June 1990. RESNA Press.

31. T. Vlachos and A. Constantinides. Graph theoretical approach to colour picture segmentation. In *IEEE Part 1*, pages 36–45, 1993.

32. L. Wolff and E. Angelopoulou. Three-dimensional stereo by photometric ratios. *Journal of the Optical Society of America: Optics and Image Science and Vision*, 11(11):3069–3078, November 1994.

33. A. Zeelenberg. Domestic use of a training robot-manipulator by children with muscular dystrophy. In R. Foulds, editor, *Interactive Robotic Aids-One Option for Independent Living: An International Perspective*, Monograph 37, pages 29–33. World Rehabilitation Fund, 1986.

Personal Adaptive Mobility Aid for the Infirm and Elderly Blind

Gerard Lacey, Shane Mac Namara, and Kenneth M. Dawson-Howe

Computer Science Department,
School of Engineering,
Trinity College Dublin,
Dublin 2,
Ireland.

Abstract. People with both visual and mobility impairments have great difficulty using conventional mobility aids for the blind. As a consequence they have little opportunity to take exercise without assistance from a carer. The combination of visual and mobility impairments occurs most often among the elderly. In this paper we examine the issues related to mobility for the blind and pay particular attention to the needs of the elderly or frail. We overview current mobility aids and detail some of the research in this area. We then describe our robot mobility aid, PAM-AID, that aims to provide both physical support during walking and obstacle avoidance. We examine factors that are relevant to the operation of PAM-AID and describe some initial user trials. Finally we describe the current status of the project and indicate its future direction.

1   Introduction

The opportunity for independent mobility is a major factor affecting the quality of life of all people. Frailty when combined with a visual impairment has a devastating effect on the ability of the elderly to move around independently. This often results in their becoming bed-ridden "for their own safety." In Europe over 65% of all blind people are over 70 years of age [13]; therefore there is a real need for a device to improve the independent mobility of the frail visually impaired. The elderly and infirm blind are excluded, by virtue of their frailty, from using the conventional mobility aids such as long canes and guide dogs. Consequently the elderly visual impaired are heavily dependent on carers for personal mobility. This level of carer involvement is often beyond the resources of a family or residential care facility and the person is forced into a sedentary lifestyle. A sedentary lifestyle accelerates the degeneration of the cardio-pulmonary system and in, addition, the increased isolation and dependence can lead to severe psychological problems.

The PAM-AID project aims to build a mobility aid for the infirm blind which will provide both a physical support for walking and navigational intelligence. The objective is to allow users to retain their personal autonomy and take independent exercise. In this research we have attempted to examine the needs

V. O. Mittal et al. (Eds.): Assistive Technology and AI, LNAI 1458, pp. 211–220, 1998.
© Springer-Verlag Berlin Heidelberg 1998

of potential users and we have been aided in this by the staff of the National Council for the Blind of Ireland (NCBI). Initially we examine the issues involved in mobility for the visually impaired and look at the state of the art in mobility aids for the blind. We then examine the mobility needs of the elderly and infirm and identify how these affect the design of our mobility aid. We then describe the design of the PAM-AID robot and report on user trials of the prototype.

2   Mobility and Navigation for the Visually Impaired

In this section we examine the currently available mobility aids and identify how they are used. We will also try to identify their limitations particularly in the case of the elderly and infirm.

2.1   The Long Cane

By far the most common mobility aid for the visually impaired is the long cane. At its most simplistic the cane is swept from left to right, synchronized with the stride of the user. The synchronization is such that the cane sweeps the space in front of the next stride. The length of the cane is the distance from the base of the sternum to the ground, thus the blind person is given approximately one stride preview of the terrain directly ahead. If an obstacle is detected the cane user must be able to react quickly to avoid a collision.

The limitations of the long cane are that the entire space through which the body moves is not scanned. Of particular importance is the fact that overhanging obstacles such as the rear of parked trucks and holes in the ground cannot be detected reliably. There can however be a high degree of stress associated with cane use due to the short preview of the terrain and the limited amount information it provides.

2.2   Guide Dogs

The other most commonly used mobility aid for the blind is the guide dog. The typical guide dog begins training at 2 years of age and has a working life of roughly nine years. Guide dogs cost approximately $16,000 to train and about $30 per month to maintain.

Contrary to the popular imagination guide dogs are not suitable as a mobility aid for all blind people. The blind person's visual impairment must be so severe as to prevent their anticipation of stops or turns before they receive this information from the dog. If the guide dog user could anticipate such events the dog would not have the opportunity to put its training into practice and without sufficient reinforcement may no longer function effectively. In addition the training process is physically strenuous and the users must have good co-ordination and balance. A typical guide dog can walk at a speed of 5 km per hour therefore the user must have an active lifestyle to provide the dog with sufficient exercise and reinforcement. The dog must be given constant correction if it does not perform correctly except in the case of *intelligent disobedience*. This is where the dog disobeys the command if it would expose the person to danger and is particularly important for crossing roads.

2.3   Walking Aids

Walking aids are used by persons with a balance or weight bearing problem. Visually impaired people do use these devices. However, due to the difficulty in sensing the environment their use is limited to those with some level of remaining vision. Although there are many different models of walking frames they fall into three distinct categories, walking frames, reciprocal walking frames and rollators.

- **Walking frames**, sometimes called "Zimmer" frames, are designed to provide a larger base of support to a person with lower limb weakness. The frame is used by lifting, placing it forwards, bearing weight through the grips and taking two strides to the center of the frame.
- **Reciprocal Frames** are similar to walking frames except that the frame is hinged on either side allowing the sides of the frame to be moved alternately.

- **Rollators** or "Strollers" are walking frames with wheels attached.

2.4   Electronic Travel Aids (ETAs)

Even though the long cane is a very cheap and reliable mobility aid it does have the drawback that all the space through which the body travels is not scanned. This leaves the upper body particularly vulnerable to collisions with overhanging obstacles or with other people. This deficit of the long cane has prompted much research into electronic travel aids (ETAs). Several reviews have been done by Nye & Bliss in [11], Boyce [3] and in [14] Farmer reviews mobility devices in depth. Some devices described by Farmer are the Wheeled Cane [10], the   C5 Laser cane [5] and the Sonicguide [8]. The wheeled cane fitted a long cane with a wheel at the bottom, a tactile compass, sonar, optical and tactile sensors. The C1 - C5 laser canes used triangulated laser diodes to detect drop-offs and head height obstacles. Output was by means of tactile and tonal output. The Sonicguide was a head mounted sonar sensor which provided binaural feedback. More recently this sensor has been developed into the KASPA sensor [1] which has been used in object and texture discrimination.

Robot ETAs for the blind have been developed by Tachi [7], Mori [6] and in recent work by Borenstein and Ulrich [2]. In the first two examples the researchers built large vision based robots to act as guide dogs. In the latter case the authors have developed a ETA by attaching a small robot to the end of a long cane. The robot uses a combination of sonar, a flux gate compass and odometry to lead the blind user around obstacles.

---

[1] www.sonicvision.co.nz/bat

Personal ETAs are not used by the majority of blind users, primarily due to the excessive cost, poor user interfaces or poor cosmetic design. If a mobility aid is to be successful the device must provide the user with a great deal more information about the environment than the long cane. It must also present this information in a manner that does not occlude the remaining senses. For example requiring the user to wear a pair of headphones would exclude noises from the environment. The device must be affordable, robust and not draw undue attention to the user's blindness. This is a difficult specification to achieve as emphasized by the continued preference for long canes and guide dogs by the majority of blind people.

3   Technology and the Elderly

In designing technology specifically for the elderly we need to address the relationship between elderly and technical aids. Fernie [4] reviews assistive devices for the elderly and their affect on their quality of life. He focuses attention on the need to retain the ability of the individual to make choices. Wellford in [1] reports that the speed and accuracy of elderly people for simple motor tasks is quite good but this deteriorates rapidly as the task complexity increases. This is particularly true if there is a extended time between the stimulus and the taking of the corresponding action. In general, where possible, the elderly shift their concentration from speed to accuracy in an attempt to maximize the use of limited physical resources.

Kay in [1] examines learning and the effects of aging. Short term memory is very dependent on the speed of perception and thus a deterioration in perceptual abilities will produce a consequent deterioration in short term memory. Learning in older people consists of the modification of earlier experiences as opposed to learning from new stimuli. This consists of a process of adapting the previous routine to the new task and features the continuous repetition of small errors.

Among the elderly, motivation for learning is much reduced as the acquisition of a new skill may not seen to be worth the effort given the limited life expectancy. Karlsson in [9] notes that usability or "perceived ease of use" is not the limiting factor in the adoption of new technology by elderly people. She shows that "perceived usefulness" is the prime factor in the adoption of a new technology as it is directly related to the users motivation. Perceived usefulness is influenced by information and is sustained by the evaluation of "service quality" parameters. Perceived ease of use on the other hand influences the adoption of new subsystems technology and is in turn influenced by hardware and software design, user experiences and by training and support. Introducing new technology into the domestic area affects that environment and this must be considered when assessing the design of the system.

4   Personal Adaptive Mobility Aid (PAM-AID)

In previous sections we have reviewed aids for the frail and visually impaired. We have also considered the users needs and factors influencing the design and adoption of a technical aid for the elderly. From this investigation we developed a specification for a mobility aid which we call the "Personal, Adaptive Mobility AID" or PAM-AID.

This research aims to build a robot mobility aid that provides both physical support and obstacle avoidance. In this work we are trying to provide limited independent mobility to a group of people who would otherwise be bed-ridden. We are not attempting to build a robotic guide dog which will work in all environments and for all people. We try to support the user's remaining abilities by increasing their confidence to take independent exercise. We do not aim to remove the necessary human contact involved in the care of the elderly; however, we hope to facilitate the greater independence of the person within a caring environment.

Fig. 1. The PAM-AID concept prototype

The aim of the concept prototype was to investigate the overall feasibility of the project by providing a focus for discussion between the authors and the representatives of the user group, the National Council for the Blind of Ireland. Our design goal at the outset was to keep the basic robot as simple as possible to facilitate user acceptance, low costs and reliability. Early investigation of the user needs highlighted that a wide variety of user interface configurations would be required to meet the needs and preferences of individual users. In particular, adaptations to cope with hearing impairment and arthritic complaints had to be considered. The robot base used for the concept prototype, shown in Fig. 1, was a Labmate robot base. The robot was fitted with a handrail to provide physical support and a joystick which indicated the users intentions. The sensors used were Polaroid sonar sensors, infra-red proximity switches and bumpers.

4.1   User Interface

The user input device was a joystick with a single switch that was mounted on the handrail. Feedback to the user was provided by means of tonal and/or recorded voice messages from the PC controller. The audio feedback played two roles:

- Command Confirmation
  When operating under direct human control the voice messages relayed the current direction indicated by the joystick.
- Warning of mode change
  When operating in wall following mode or if an obstacle was detected the robot issued a warning that it was about to change direction or stop.

The robot operated in two modes: direct human control and wall following mode. The user selected between these modes by holding down the switch on the joystick. As long as the switch was held down the robot approached the nearest wall and began following it. In direct human control the user indicated their desired direction via the joystick. The robot adopted this direction at a gentle speed only stopping if an obstacle was encountered.

4.2   Control System

In this application it is difficult to separate the user interface from the control system as this is a typical example of a *human in the loop control*. The speed and manner of mode/direction switching determines a great deal about the user's experience of the robot. If the robot responds too quickly it can misinterpret the user's intention. Typically a debounce delay of 300ms was required before a command would be accepted; in addition, acceleration had to be slow to prevent the position indicated on the joystick being affected by the robot motion. The controller adopted for the concept prototype involves complete directional control by the user with the robot only providing direction assistance via speech feedback and stopping before dangerous situations occur. Control over direction can be swapped between the user and the robot by the user depressing a switch.

The control system was implemented as a subsumption architecture [12]. At the lowest layer in the hierarchy was the **Avoid Collision** behavior. It detected the presence of an obstacle, issued a warning message and slowed the robot and eventually stopped before a collision could occur. At the next level in the hierarchy there were two parallel behaviors, **Wall Following** and **Direct Joystick Control**. Arbitration between these behaviors was achieved by the user selecting, via a switch input, which behavior would have highest priority.

4.3   Evaluation of the Concept Prototype

The concept prototype was evaluated in the laboratory by representatives from the National Council for the Blind of Ireland and by researchers from the Sensory Disabilities Research Unit, University of Hertfordshire, UK. The evaluators were able bodied; however, all were involved in providing mobility training to the visually impaired or in research related to the impact of visual impairment. In a separate exercise the opinion of potential users to the PAM-AID concept was sought by both the authors and the evaluators.

Concerns over the safety of the device were expressed by both carers and users. The most important factor was the detection of descending stairs. In the words of one mobility expert "If the device fails to detect descending stairs it will be useless." The evaluators and users were concerned that the device must be extremely responsive to user input, i.e., not drag the users after it or exert any force on them which might upset their balance.

A great deal of attention was paid to the user interface of the device. Many of the preferences expressed by different users were contradictory confirming the requirement for customization of the user interface. A typical example was the preference by some people for voice control of the robot while others prefer switch based input. Cultural and personal differences also produced a wide spectrum of responses to the whole concept of a robot mobility aid. Some users were delighted at the prospect of regaining their independence while others would "prefer to crawl" rather than use a walking frame.

The process of introducing a robot aid into the lives of potential users requires a flexible user interface and control system. Initially the users would prefer to have only limited control over such parameters as speed, acceleration, and user interface configuration, however as they become more familiar with the device they would like to have increasing control over the various parameters of the robot. A typical example would be the disabling of voice feedback in a certain situations or changing the robot speed on command.

5   The PAM-AID Rapid Prototype

Following the evaluation of the concept prototype and the user needs survey the PAM-AID Rapid Prototype was designed. The users had expressed a preference for handles rather than a handrail, also the handles had to be height adjustable. We chose to build our Rapid Prototype around a commercially available rollator. This was fitted with a custom built drive system as shown in Fig. 2.

Two types of user interface were developed for the user trials, instrumented handles and finger switches. The instrumented handles fitted with two micro switches at the limits of a 5 degree pivot. The micro switches detect if the handle is being pushed forward, pulled back, or in neutral. The finger switches consist of four switches for forward, reverse, left and right turn. In addition to these two options the user interface also consists of two finger switches, one an enable switch which must be pressed for the robot to move and a second which invokes the **Wall Following** behavior.

Fig. 2. PAM-AID Rapid Prototype

5.1 Field Trials of Rapid Prototype

In June 1997 the Rapid Prototype was tested in two residential homes for the elderly visually impaired in the UK. Eight subjects tested the device using the two different configurations of the user interface. Photographs of the user trials are shown in Fig. 3.

6   Future Work

The current PAM-AID research project is due to finish in July 1999[2] with the completion of a robot which can be used for extended user trials. The future direction of the research will focus on user interface options, multi-sensor fusion and the development of a shared control system. Currently we are integrating a voice control system into the user interface. This system will be speaker dependent to prevent other people from causing the robot to move. The high level nature of voice commands and their low frequency represent a challenge in the design of the control system. When using the switch input or joystick input commands can be given to the robot several times a second whereas when using voice input this is not possible. Due to these limitations and a need to reduce the cognitive load on the user we are developing a shared control system. The aim of the shared control system will be to determine the users high level goals and the control system will produce a plan of action for the robot.

---

[2] Regular updates will be posted on the web at www.cs.tcd.ie/PAMAID

Fig. 3. User Trials of PAM-AID Rapid Prototype

A major factor in the operation of the device will be docking the robot and user with chairs, beds, etc. This will require some limited feature detection and path planning capabilities. Currently the robot uses sonar and bumper sensors. To expand its functionality we are integrating a laser scanner for feature detection and infra red sensors to detect drop-offs. We aim to use a probabilistic data fusion protocol to provide the user with information of the presence and location of obstacles and features such as doors, chairs, etc.

7 Conclusions

This work is seen as part of a long term effort to apply Artificial Intelligence and Robot Technology to the needs of the wider community. We have chosen a well focused project such as PAM-AID as it represents both a concrete need and a significant challenge. The needs of the infirm blind and visually impaired are quite different from those of the able-bodied blind. This manifests itself in the need to combine both a walking support and a mobility device. We are in the early stages of this work and are concentrating on developing the user interface and control systems required to provide a reliable mobility aid in a dynamic environment. We aim to develop a modular robot design in which complex tasks and user interfaces can be customized to meet the needs of individual users.

By placing a human being at the center of the design of the device we have had to consider several interesting research issues. The primary one is the users relationship with the device. The short term memory problems of the elderly and the likelihood of their being some cognitive dysfunction constrain it to being as simple and intuitive as possible. The provision of feedback on the environment must be based on the needs of the user (reassurance, information) and the needs of the robot (user safety). The modalities of this feedback must be flexible to cope with a range of user preferences.

The research contributes to general research in AI in that it focuses attention on how humans represent and use environmental information. The lessons learned in developing applications for the disabled will contribute to other AI domains such as tele-operation, sensing, planning and control.

8 Acknowledgments

We would like to acknowledge the financial support of Irish National Rehabilitation Board (NRB) and the Computer Science Department, Trinity College Dublin. From January 1997 this project has been funded under the EU Telematics Applications Program for the Integration of Elderly and Disabled people, Project 3210 PAM-AID.

References

1. J. Birren. *Handbook of aging and the individual.* Chicago University Press, 1959.
2. J. Borenstein and I. Ulrich. The guide cane - a computerized travel aid for the active guidance of blind pedestrians. In *IEEE International Conference on Robotics and Automation*, April 1997.
3. K. Boyce. Independent locomotion by blind pedestrians. Master's thesis, Department of Computer Science, Trinity College Dublin, 1991.
4. G. Fernie. Assistive devices, robotics and quality of life in the frail elderly. In Rowe J.C. Birren J.E., Lubben J.E. and Deutchman DE, editors, *The Concept and Measurement of the Quality of life in the Frail Elderly.* Academic Press, 1991.
5. J.M. Benjamin, Jr. The new c-5 laser cane for the blind. In *Proceedings of the 1973 Carahan conference on electronic prosthetics*, pages 77–82. University of Kentucky Bulletin 104, November 1973.
6. M.H. Yasutomi, S. Mogodam, C.N. Nishikawa, K. Yamaguchi and S. Kotani. Guide dog robot harunobu-5. In *Proceedings of SPIE Mobile Robots VII Conference*, 1992.
7. S. Tachi and K. Komoriya. Guide dog robot. In H. Hanafusa and H. Inoue, eds., *Second International Symposium of Robotic Research*, pages 333–340. MIT Press, 1985.
8. L. Kay. A sonar aid to enhance the spatial perception of the blind: engineering design and evaluation. *Radio and Electronic Engineer*, 44(11):605–627, November 1974.
9. M.A. Karlsson. Elderly and new technology - on the introduction of new technology into everyday life. In I. Placencia Porrero and R. Puig de la Bellacasa, eds., *The European Context for Assistive Technology*, pages 78–81. IOS press, April 1995.
10. H.L. Martin. New type of cane for the blind (wheeled cane). Technical Report MFS-21120, George C. Marshall Space Flight Center, NASA, October 1970.
11. P.W. Nye and J.C. Bliss. Sensory aids for the blind : a challenging problem with lessons for the future. *Proceedings of the IEEE*, 58(12):1878–1898, 1970.
12. R.A. Brooks. A robust layered control system for a mobile robot. *IEEE transactions on Robotics and Automation*, 2, April 1986.
13. B. Richards. *Results of a survey carried out by the Hampshire Association for the care of the Blind.* Hampshire Association for the Care of the Blind, Hampshire, SO5 6BL, UK, 1993.
14. R.L. Welsh and B.B. Blasch, eds. *Foundations of Orientations and Mobility.* American Foundation for the Blind, 15 West 16th Street, New York, N.Y. 10011, 1987.

HITOMI: Design and Development of a Robotic Travel Aid

Hideo Mori, Shinji Kotani, and Noriaki Kiyohiro

Department of Electrical Engineering and Computer Science
Yamanashi University
Kofu Yamanashi, 400
Japan
{forest,kotani,kiyohiro}@koihime.esi.yamanashi.ac.jp

Abstract. A *Robotic Travel Aid* (RoTA) is a motorized wheelchair e-quipped with vision, sonar, and tactile sensors and a map database system. A RoTA can provide a visually impaired user assistance with orientation and obstacle avoidance, as well as information about their present location, landmarks, and the route being followed. In this paper we describe HITOMI, an implementation of the RoTA concept that can guide a visually impaired user along a road with lane marks or along a sidewalk while avoiding obstacles.

1 Introduction

A number of electronic mobility aids [15,14,8] have been developed for the visually impaired. Among them the Mowat sensor and Sonic Guide have been available for 20 years, but have not become widespread. Why are they not commonly used by blind people? Jansson [6] has suggested that it is because most aids give information about the environment only a few meters ahead that can easily be obtained by using a long cane. Another reason, we believe, is that the sound by which aids communicate with users disturbs their echo-location and do not adequately inform the user of the environment.

The guide dog is the best travel aid, but it is difficult to train enough guide dogs. The number of guide dogs in the world (e.g. 8,000–10,000 in USA, 4,000 in UK, 730 in Japan) illustrate this difficulty. In Japan, a shortage of training staff and budget (it requires about 25,000 dollars (US) to train each guide dog) makes it difficult for the guide dog to become widespread. The Mechanical Engineering Laboratory of Japan started project MELDOG to develop a robotic guide dog in 1977 [16]. MELDOG used CCD sensors to detect bar-code-like landmarks fastened on the road. However, this research project ended after seven years and the robot was never used by the blind. Recently, Vision Guided Vehicles and Electronic Travel Aids (ETAs) have been proposed by several researchers (e.g., [3,11]). Like our system, these follow a marked lane, avoid obstacles, and aim to provide full navigation and mobility capability for blind and partially sighted people.

V. O. Mittal et al. (Eds.): Assistive Technology and AI, LNAI 1458, pp. 221–234, 1998.
© Springer-Verlag Berlin Heidelberg 1998

We have proposed a behavior based locomotion strategy called *Sign-Pattern Based Stereotyped Motion* [13]. Recently, automobile navigation systems have been produced commercially by the electronics and automobile industries; since 1990 we have combined these technologies to develop *Robotic Travel Aids* (RoTAs) [9]. In this paper we describe the RoTA HITOMI ("pupil" in Japanese). Photographs of HITOMI are shown in Figures 1 and 2.

Fig. 1. Robotic Travel Aid HITOMI

2   RoTA Requirements

A RoTA is not a substitute for the guide dog. Rather, it is an advancement of ETA. Required functions of a RoTA include:

Target of RoTA

The required functions of a RoTA are different depending on the level of visual disability, the age the user lost their sight, and whether or not the user is hearing impaired. In general, the older a person is when they lost their sight, the more difficult it is to train them to use echo-location. Furthermore, for an older person, it is very difficult to memorize and recall entire routes. HITOMI is designed for those who lost their sight later in life and have difficulty remembering routes.

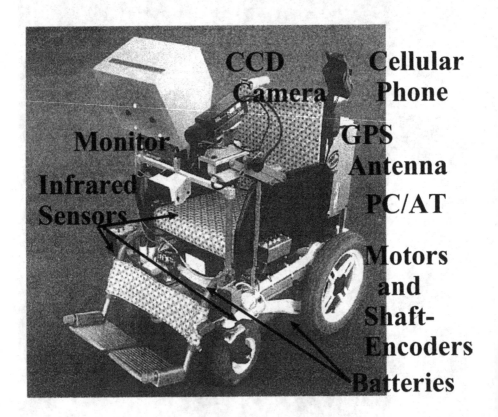

Fig. 2. Hardware Configurations of HITOMI (new).

Size of RoTA

To utilize a video camera as a sensor, a RoTA must be large enough to stabilize the video image. A motorized wheel chair is used as the under-carriage of HITOMI. It is 1,170 (Length), 700 (Width), 1500 (Height) millimeters in size and 80 kilograms in weight. The wheelchair is big and heavy enough for the blind person to walk while holding the handle bar.

Sensors of RoTA

A RoTA requires information about orientation and mobility. Orientation information required to reach the goal is obtained by the use of vision to detect passages without obstacles. Mobility information is required by blind people to control walking. The DGPS is to determine the robot's initial position and orientation. HITOMI uses sensors to get information for orientation and mobility. Table 1 shows multiple sensors of HITOMI.

Table 1. Multiple Sensors of RoTA

Sensor	Range [m]	Objects
Vision	3 – 50	Elongated features, free space, vehicles,pedestrians
Sonar	less than 3	Wall-like obstacles
Tactile	less than 0.3	Depressions, Stairs
DGPS		Decision of an Initial Position and Orientation

Information about orientation can be obtained by sensing the passage ahead within 5 meters. Only vision can provide an adequate range of sensing. We believe that monocular and monochromatic vision is sufficient for orientation. Vision is also useful to detect obstacles such as vehicles and pedestrians. However, it can not detect wall-like obstacles with a homogeneous surface. Sonar is useful to detect wall-like obstacles, but its effective range is limited to 3 meters in an outdoor environment.

Sign Pattern Based Stereotyped Motion

We have previously proposed a behavior-based action strategy based on the idea of a *Stereo Typed Motion* (STM) [12]. An STM is a fixed action pattern that makes the robot perform a specific skilled action. We assume that five STM's—*moving along, moving toward, moving for sighting, following a person* and *moving along a wall*—are adequate to follow any route from start to finish. Complex actions such as obstacle avoidance can be defined as chains of these STM's.

A pattern of features of the environment used to initiate or modify an STM to fit the environment is called a *sign pattern* (SP). The use of STM's is different from subsumption architecture [1] which does not use SPs. Our use of STMs is a goal-oriented action and can perform a mission or task, but subsumption architecture is reactive and cannot perform a mission.

The advantage of using SP-based STMs are 1) The robot can move even when the information about the future part of the passage is incomplete. For instance, if the next part of a route is invisible because it is beyond a corner, the robot can turn the corner by a chain of STMs that includes collision avoidance for an obstacle that might suddenly appear, and 2) the reaction time to a SP is very short, as it does not need motion planning.

Moving Along and Moving Toward

In his psychological study, Jansson [6] defined two kinds of perceptual guiding actions, *walking along* and *walking toward*. Applying these to RoTA we define two STMs; *moving along* and *moving toward*.

Moving along is defined as an STM that consists of two cooperative actions of the under-carriage and video camera systems. It moves along an elongated SP keeping the distance from SP constant, changing the camera direction to keep the SP in the center of the video image. The elongated SP may include a lane mark of a road, the edge of a sidewalk, fences or the boundary of a campus path.

Moving toward is also defined as an STM that consists of the cooperative action of the two systems to move toward a goal. The RoTA has to search for the SP of a goal in its video image. An SP of a goal includes only a crosswalk mark at present, but in the future it will include entrances of buildings and stairs.

Obstacle Avoidance

When RoTA moves along a road or sidewalk most critical obstacles are vehicles and bicycles. The RoTA's obstacle avoidance is carried out through four tasks: *moving along* in which obstacles are found, *moving for sighting* finds the right-hand side of the obstacle, *moving along wall* which passes by the obstacle, and again *moving along*. These are illustrated in Figure 3.

Map-Based Guidance

The digital map system of RoTA gives two kinds of map information, one is for the RoTA itself and another is for the impaired person.

The map is a *full metric model* [10] of an environment. Figure 4 is a schematic representation of our environment map. Paths are expressed as networks which are specified with fixed coordinates. Networks consist of nodes and arcs. Locations of landmarks and sign patterns are specified with coordinates, and landmarks may have features. A network's data is similar to the road information for digital car navigation maps. A landmark's data and the sign patterns data are added by human beings.

The map information for the user is used to let him know the landmarks of the present location and the route represented by a command list such as *go straight* or *turn right*.

Crossing an Intersection

A major problem for the visually impaired while walking outdoors is safely crossing road intersections. A RoTA can find an intersection if there is crosswalk mark on the road by detecting it visually.

Semi-Automatic Navigation

While an SP is detected visually, the impaired person may follow a RoTA which is *moving along* or *moving toward*. However, when the SP disappears RoTA stops, refers to the map system and makes an inference as to why the SP has disappeared.

The inference of the RoTA may be *we have reached the end of the SP* or *we have probably met an obstacle*. The user can test the environment through his auditory and tactile senses, and understand the situation through his knowledge of environment, traffic, weather and time. After the RoTA makes a list of possible STMs, the user can select one to be performed.

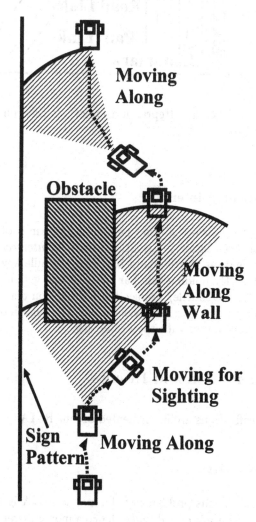

Fig. 3. The Sign Pattern Based Stereotyped Action for an Obstacle Avoidance.

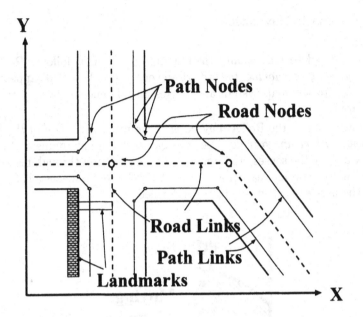

Fig. 4. A Schematic Representation of our Environment Model.

The Blind-Oriented Interface

The RoTA has to inform the blind person of four kinds of information: *mobility* information. *orientation* information, *obstacle/intersection* information and *map-based* information. A command bar with a braille key is fixed on the rear part of the robot. By holding this bar the user can get the mobility and orientation information. Obstacle/intersection information detected by the RoTA is issued as warning and alarm messages through the voice interface. The user tells the RoTA the destination with the braille key.

3 Implementation and Results

The above specifications are implemented on our RoTA.

3.1 Under-Carriage

One of the most serious problems of the mobile RoTA is that it cannot go up or down stairs. The motorized wheelchair cannot go over a step more than 3 centimeters in height.

3.2   Camera Platform

The major problems of the camera platform include reducing the electric power consumption and decreasing the vibration of the video image during locomotion. We originally used servo systems intended for factory automation, but they had high power consumption and were very large. We now use commercially available servo systems developed for model aircraft. These have only 20% the size and power consumption compared to industrial servo systems. Their direction accuracy is 0.5 degree in pan and tilt angle. To decrease vibration, the camera platform is supported on the RoTA by shock absorbing rubber.

3.3   Vision System

One of the most difficult problems of the vision system is reducing the electric power consumption while keeping the image processing time short. At present we use one board image processing system based on an MC-68040 processor which has an image buffer and 4 megabytes of main memory which is accessible through cache memory. It consumes 60 watts.

We have developed software routines which can process a scene within two or three frames (66 or 99 msec). We apply dynamic vision as described by Graefe [5].

3.4   Road and Lane Mark Detection

Understanding of the road and sidewalk image begins with lane mark detection. To detect the lane mark as an elongated SP, the monocular video image is binarized by the mode method based on a gray-level histogram. The SP is detected every two frames (66 msec). We use Kalman filtering to eliminate false data from the SP.

3.5   Crosswalk Mark Detection

As a RoTA approaches a road intersection it detects it using map matching, and inserts a crosswalk mark searching process into every ten cycles of the lane mark detection process [17]. When the mark is found, it is followed until the RoTA is 3 meters from it. In its searching and following process the road image is binarized, and horizontal and vertical projection are performed on the image. By analyzing the two projections the mark is identified. Since some obstacles momentarily show almost the same projections as the mark, the RoTA sometimes mistakes the obstacle for the mark, but this error can be corrected by checking the projection in the successive frames.

3.6   Vehicle Detection

Vehicles are the most troublesome obstacles, whether they are moving or stationary. We have proposed a simple useful vehicle detection algorithm based on the fact that the space beneath a vehicle is relatively dark [2]. Although the space below a vehicle and shaded areas appear to have the same luminance, this is an illusion of brightness constancy. Our algorithm is easy to implement, and right and left edges of the vehicle are also used for vehicle identification. An example of an intensity curve in a window which is located beneath a vehicle is shown in Figure 5. We tested this method in four traffic scenes consisting of 1) a partly shaded road, 2) an entirely shaded road, 3) a non-shaded road, and 4) a road in cloudy conditions. More than 97% of vehicles were successfully detected (Table 2).

Table 2. Results of Vehicle Detection

Weather	Shadow	Number	Success [%]	Failure[%]	
				Negative	Positive
Fine	Partly	294	92.0	7.0	1.0
Fine	Entirely	191	96.8	3.2	0.0
Fine	None	272	97.8	2.2	0.0
Cloudy	None	405	98.4	1.3	0.3

3.7 Pedestrian Detection

Pedestrian detection is very important when moving along a sidewalk. We have proposed a rhythm model to detect and follow pedestrians [18]. The model is based on the observation that when a person walks their volume changes rhythmically in area, width, and height. The volume is specified by the mean and standard deviation of walking frequency. Advantages of the rhythm model include:

- Total image processing time is shorter than for any other model because few non structured features are used in processing.
- Rhythm does not vary for different distances between person and observer.
- Rhythm is independent of illumination and therefore robust with respect to time and weather changes.
- Rhythm is easy to detect when a person is wearing clothes.

A disadvantage of the rhythm model is that it can be intentionally deceived.
Pedestrian detection by the rhythm model is composed of four processes:

1. *Moving Object Detection* We applied a method based on subtraction of successive frames. This is why the method is very fast. In order to get the bottom position of the object region, the vertical and horizontal projection of the intensity are calculated as shown in Figure 6.
2. *Tracking a Moving Object* There are two methods to implement tracking. The first is to seek the unique shape of the object [4]. The second method

is based on the kinematics of the object [7]. Our method is based on the latter. An *Extended Kalman Filter* (EKF) is used to get the estimated state vector and predicted measurement. The estimated state vector is the result of the incremental procedure of EKF. The predicted measurement points to the center of the tracking window at the next time step.

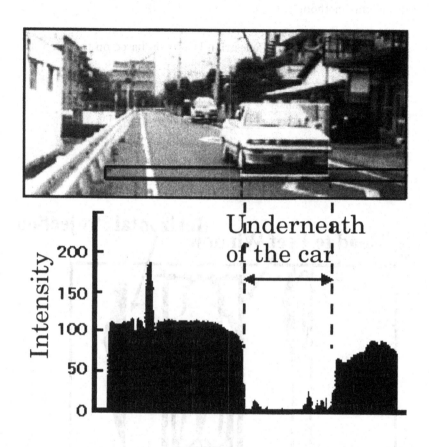

Fig. 5. The Sign Pattern of Vehicles and Intensity Histogram.

3. *Checking the Validity of Measurement* The pedestrian may be hidden by other moving objects and noise may lessen the validity of measurement. These facts reduce the reliability of measurement and will cause tracking to end in failure. For these reasons, we should make sure that observation is reliable. If the observation does not satisfy the conditions, the observed position is judged to be wrong and should be replaced with the predicted position.

4. *Finding a Pedestrian Based on the Rhythm* The rhythm of walking is caused by a two-stage bipedal action: first, a pedestrian stands still for a relatively long time on both feet; second, one of the feet steps forward rather swiftly.

This is clearly seen in Figure 7. as the periodic intensity change of the sub-tracted image data around the feet. Figure 8 shows the time series of the area. Figure 9 shows the rhythm of walking is 1 second in frequency. This method works well when the robot stops and looks at pedestrians at a distance of 7–30 meters through the video camera. Table 3 shows the success rates of this method.

Table 3. Results of the Pedestrian Detection based on the Rhythm

Objects	Sampling No.	Success [%]	Failure [%]
Pedestrian	533	94.0	6.0
No Pedestrian	109	95.4	4.6

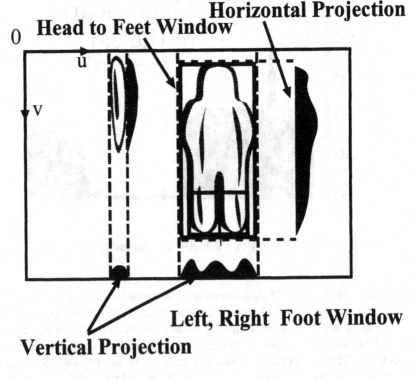

Fig. 6. A Window Setting for Detection of Intensity Changing.

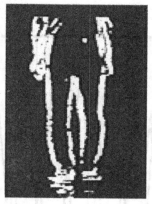

Fig. 7. Subtracted Images at the Feet of a Pedestrian. The Contrast was Enhanced for Clarity. from Left to Right: 0.00 seconds, 0.27 second, 0.53 seconds

4   Concluding Remarks

We have been developing HITOMI, a robotic travel aid, which guides the visually impaired along roads and paths. HITOMI is a small mobile robot which utilizes a motorized wheel chair as its under-carriage. A vision system is equipped to detect the road, vehicles and pedestrians. A sonar system is used to detect walls and other obstacles which the vision system cannot detect. A portable digital map system is used to give the under-carriage a sequence of commands to follow routes from start to finish. It also gives the vision system detection parameters of sign patterns and landmarks along the route. The digital map includes the names of intersections and buildings. The user can ask where he is and the map system replies through a synthesized voice. A command bar is attached at the rear part of HITOMI. The user stands behind the RoTA and follows by grasping the command bar. He can get mobility and orientation information through the motion of HITOMI.

The success rate of vehicle and pedestrian detection is between 92% and 94%. To avoid accidents, HITOMI uses semi-automatic navigation. When HITOMI senses an environmental change, it infers its cause and tell the user its inference and the next plan of motion using a synthesized voice. The user confirms the inference using his residual senses and permits HITOMI to perform the plan, or makes it wait for his permission. Generally speaking, the impaired person does not want to have to obey completely what the robot commands. By the active use of his residual senses, his independence of life will be promoted.

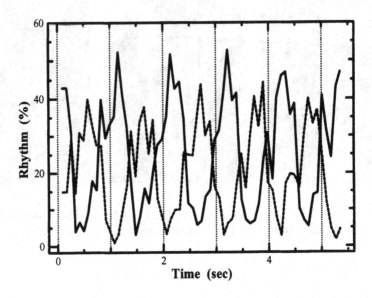

Fig. 8. Time Series Data of the Periodic Intensity Change; Solid Line: a Left Foot, Dotted Line: a Right Foot.

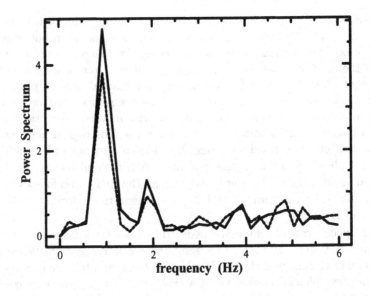

Fig. 9. A Power Spectrum of a Pedestrian; Solid Line: a Left Foot, Dotted Line: a Right Foot.

References

1. R. A. Brooks. A robust layered control system for a mobile robot. *IEEE Trans. Robotics and Automation*, 2(1):14–23, 1986.
2. N. M. Charkari and H. Mori. A new approach for real time moving vehicle detection. In *IROS'93*, pages 272–278, 1993.
3. The MoBIC Consortium. Mobility of blind and elderly people interacting with computers. Published by Royal National Institute for the Blind on behalf of the MoBIC Consortium, April 1997.
4. A. L. Gilbert, M. G. Giles, G. M. Flachs, R. B. Rogers, and Y. H. U. A real-time video tracking system. *IEEE Trans. Pattern Anal. Mach. Intell. PAMI-2*, 1:49–56, 1980.
5. V. Graefe. Dynamic vision system for autonomous mobile robot. In *IROS'89*, pages 12–23, 1989.
6. G. Jansson. Non-visual guidance of walking. In R. Warren and A.H. Wertheim, editors, *Perception and Control of Self-Motion*, pages 507–521. Lawrence Erlbaum Associate, 1990.
7. G. R. Legters Jr. and T. Y. Young. A mathematical model for computer image tracking. *IEEE Trans. Pattern Anal. Mach. Intell. PAMI-4*, 6:583–594, 1982.
8. L. Kay. An ultrasonic sensing probe as a mobility aid for the blind. *Ultrasonics*, 2(53–59), 1964.
9. S. Kotani, H. Mori, and N. Kiyohiro. Development of the robotic travel aid hitomi. *Robotics and Autonomous Systems*, 17:119–128, 1996.
10. D. Lee. *The Map-Building and Exploration Strategies of a Simple Sonar-Equipped Mobile Robot*. Cambridge University Press, 1996.
11. N. Molton, S. Se, J. M. Brady, D. Lee, and P. Probert. Autonomous system for mobility orientation navigation and communication. Technical report, Project 1228-ASMONC, Dept of Engineering Science, University of Oxford, 1997.
12. H. Mori. A mobile robot strategy. In *The Fifth Int. Symposium of Robotic Research*, pages 162–172. The MIT Press, MA, 1990.
13. H. Mori. Guide dog robot harunobu-5: Stereotyped motion and navigation. In G. Schmidt, editor, *Information Processing in Autonomous Mobile Robots*, pages 135–149. Springer, Berlin, 1991.
14. N. Pressey. Mowat sensor. *Focus*, 3:35–39, 1977.
15. L. Russell. Travel path sounder. In *the Rootterdam Mobility Research Conference*. American Foundation for the Blind, 1965.
16. S. Tachi and K. Komoriya. Guide dog robot. In H. Hanafusa and H. Inoue, editors, *Robotics Research: The Second Int. Symposium*, pages 333–340. The MIT Press, Cambridge, 1984.
17. K. Yamaguchi and H. Mori. Real time road image processing for mobile robot. In *Int. Conf. on Advanced Mechatronics*, pages 360–364, 1993.
18. S. Yasutomi, S. Kotani, and H. Mori. Finding pedestrian by estimating temporal-frequency and spatial-period of the moving objects. *Robotics and Autonomous Systems*, 17:25–34, 1996.

NavChair: An Assistive Wheelchair Navigation System with Automatic Adaptation

Richard C. Simpson[1], Simon P. Levine[2], David A. Bell[3], Lincoln A. Jaros[4], Yoram Koren[5], and Johann Borenstein[5]

[1] Metrica Corporation
1012 Hercules, Houston
TX 77058
U.S.A.
[2] Rehabilitation Engineering Program
Depts. of Physical Medicine and Rehabilitation and Biomedical Engineering
University of Michigan, Ann Arbor, MI 48109
U.S.A.
[3] Amerigon Corporation
5462 Irwindale Avenue, Irwindale, CA 91706
U.S.A.
[4] PhaseMedix Consulting
24380 Orchard Lake Road, Suite 114
Farmington Hills, MI 48336
U.S.A.
[5] Department of Mechanical Engineering
University of Michigan, Ann Arbor, MI 48109
U.S.A.

1 Introduction

The NavChair Assistive Wheelchair Navigation System [1], shown in Figure 1, is being developed to provide mobility to those individuals who would otherwise find it difficult or impossible to use a powered wheelchair due to cognitive, perceptual, or motor impairments. The NavChair shares vehicle control decisions with the wheelchair operator regarding obstacle avoidance, door passage, maintenance of a straight path, and other aspects of wheechair navigation, to reduce the motor and cognitive requirements for operating a power wheelchair.

This chapter provides an overview of the entire NavChair system. First, the NavChair's hardware and low-level software are described, followed by a description of the navigation assistance algorithms which are employed. Next, three distinct modes of operation based on these navigation algorithms and implimented in the NavChair are presented. Finally, a method for mode selection and automatic adaptation is described.

2 System Overview

The NavChair prototype is based on a Lancer power wheelchair. The components of the NavChair system are attached to the Lancer and receive power from the

V. O. Mittal et al. (Eds.): Assistive Technology and AI, LNAI 1458, pp. 235–255, 1998.
© Springer-Verlag Berlin Heidelberg 1998

chair's batteries. As shown in Figure 2, the NavChair system consists of three units: (1) an IBM-compatible 33MHz 80486-based computer, (2) an array of 12 ultrasonic transducers mounted on the front of a standard wheelchair lap tray, and (3) an interface module which provides the necessary circuits for the system.

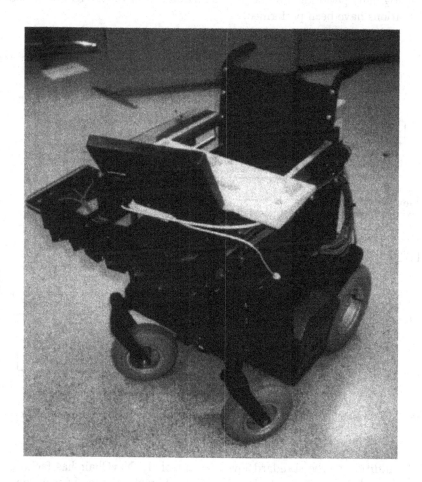

Fig. 1. The NavChair Assistive Wheelchair Navigation System

The Lancer's controller is divided into two components: (1) the joystick module, which receives input from the user via the joystick and converts it to a signal representing desired direction, and (2) the power module, which converts the output of the joystick module to a control signal for the left and right wheel motors. During operation the NavChair system interrupts the connection between the joystick module and the power module, with the user's desired trajectory (represented by input from the joystick or an alternative user interface) and the the wheelchair's immediate environment (determined by readings from the sonar sensors) used to determine the control signals sent to the power module [2]. The NavChair's software performs the filtering and smoothing operations that were originally performed by the joystick module after the navigation assistance calculations have been performed.

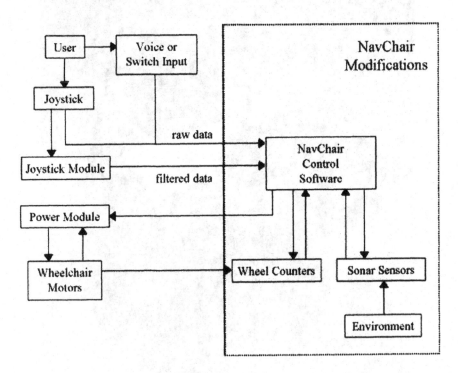

Fig. 2. Functional Diagram of The NavChair Prototype's Hardware Components [2]

In addition to the standard joystick control the NavChair has facilities for voice control. The voice control option is based on the Verbex SpeechCommander, a commercially-available continuous-speech voice recognition system that relays user commands to the NavChair via the computer's serial port. Prior to operation, users train the SpeechCommander to identify a small set of commands, a process which is typically accomplished in less than ten minutes. During operation, the user speaks a command into the SpeechCommander's microphone, worn on a headset. The SpeechCommander identifies the sound signal as one of the pre- trained commands and transmits a computer code associated with that command to the NavChair's computer. The NavChair's computer matches the signal from the SpeechCommander to a specific joystick command which is then used to steer the chair. The methods used for voice control also permit the use of discrete switches for Navchair operation.

Table 1 contains a list of the voice commands currently implemented within the NavChair. The NavChair's navigation assistance limits the commands needed to successfully complete most navigation tasks. Limiting the number of commands is desirable because it decreases the amount of time necessary to train the speech recognition system to recognize each subjects voice and the amount of time needed to teach each subject the voice control commands.

Table 1. List of Voice Commands

Command	Description
Stop	The NavChair comes to an immediate halt.
Go Forward	The NavChair begins moving at a constant speed in the direction that the chair is facing.
Go Backward	The NavChair begins moving at a constant speed in the direction opposite to that which the chair is facing.
Soft Left	The NavChair makes a small (approximately 10 degree) left turn.
Hard Left	The NavChair makes a large (approximately 20 degree) left turn.
Rotate Left	The NavChair begins rotating (in place) to the left until the operator tells it to stop or move forward.
Soft Right	The NavChair makes a small (approximately 10 degree) right turn.
Hard Right	The NavChair makes a large (approximately 20 degree) right turn.
Rotate Right	The NavChair begins rotating (in place) to the right until the operator tells it to stop or move forward.

The NavChair uses sonar sensors because of their operational simplicity and low cost. However, individual sonar readings are often erroneous. The method used to reduce these errors and create a sonar map of the chair's surroundings is called the Error Eliminating Rapid Ultrasonic Firing (EERUF) method [3]. The accuracy of the map is further enhanced by keeping track of the wheelchair's motion via wheel rotation sensors built into the Lancer's wheel motors. The result is a sonar map that is surprisingly accurate given the constraints of individual sonar sensors. The NavChair is able to accurately locate obstacles within five degrees of angular resolution relative to the center of the chair despite the fact that the resolution of an individual sonar sensor exceeds 15 degrees [4].

3   Navigation Assistance Algorithms

Two navigation assistance routines, Minimum Vector Field Histogram (MVFH and Vector Force Field (VFF), are used by the NavChair. Both stem from routines originally developed for obstacle avoidance in autonomous mobile robots. The influence of each routine on the NavChair's direction of travel at any given time is determined by the NavChair's current operating mode and immediate surroundings. This section describes the rationale behind both navigation assistance routines and gives an overview of each routine's operation.

3.1   Minimum Vector Field Histogram (MVFH)

The original obstacle avoidance technique used in the NavChair, the Vector Field Histogram method (VFH) [5,4], was originally developed for autonomous mobile robots. During development of the NavChair, it was discovered that several modifications to the original VFH method were required in order for VFH to make the transition from autonomous mobile robots to wheelchairs. One difficulty in applying an obstacle avoidance routine developed for a robot to a wheelchair is the different shapes of the two platforms. Mobile robots in general (and those VFH was originally intended for in particular) are round and omni-directional, which simplifies the calculation of trajectories and collision avoidance. While VFH has been applied to "non-point" mobile robots similar in nature to a wheelchair [6] it was determined that VFH could not support all of the desired functions (door passage in particular) while also ensuring the safety of the operator and vehicle during operation.

Another problem arose from what is considered one of the VFH method's greatest strengths, the ability to move through a crowded environment with a minimal reduction in speed. While this is acceptable for an autonomous robot, it can result in abrupt changes in direction which a wheelchair operator is likely to consider "jerky" and unpredictable behavior.

In response to these needs, the Minimal VFH (MVFH) method was developed [7,8]. The MVFH algorithm proceeds in four steps:

1. Input from the sonar sensors and wheel motion sensors is used to update a Cartesian map (referred to as the certainty grid) centered around the chair. The map is divided into small blocks, each of which contains a count of the number of times a reading has placed an object within that block. The count within each block represents a certainty value that an object is within that block, thus the more often an object is seen within a block the higher its value.
2. The certainty grid is converted into a polar histogram, centered on the vehicle, that maps obstacle density (a combined measure of the certainty of an object being within each sector of the histogram and the distance between that object and the wheelchair) versus different directions of travel.
3. A weighting function (curve w in Figure 3) is added to the polar histogram (curve h), and the direction of travel with the resulting minimal weighted

obstacle density (s) is chosen. As seen in Figure 3, the weighting function is a parabola with its minimum at the direction of travel indicated by the wheelchair's joystick position. Thus, the direction indicated by the user's input from the joystick receives the least amount of additional weight (obstacle density) and those directions furthest from the user's goal receive the most weighting, which predisposes the chair to pursue a direction close to the user's goal.

4. The wheelchair's speed is determined based on the proximity of obstacles to the projected path of the chair. This step models the rectangular shape of the wheelchair exactly when calculating the projected path, which allows the chair to approach objects closely while still maintaining the safety of the vehicle.

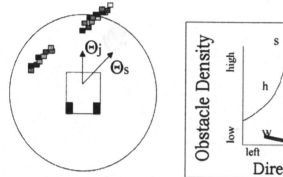
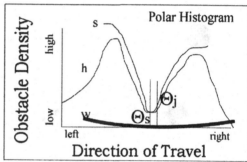

Fig. 3. MVFH Obstacle Avoidance. The left figure shows the certainty grid around the NavChair with darker shading of a cell corresdponing to a higher certainty value of an obstacle being at that location. The right figure shows the polar histogram at the same instant, where: j is the desired direction of travel, as indicated by the user with the joystick; h is the polar histogram representing obstacle densities in each possible direction of travel; w is the weighting function symmetrical about the desired direction of travel (j); s is the sum of h and w; s is the actual direction of travel selected by MVFH at the minimum of s [9].

Using MVFH, control of the chair becomes much more intuitive and responsive. Small changes in the joystick's position result in corresponding changes in the wheelchair's direction and speed of travel. Second, by modeling the exact shape of the NavChair it is possible to perform previously unmanageable tasks, such as passing through doorways. Most importantly, however, MVFH provides an adaptable level of navigation assistance. By changing the shape of the weighting function, MVFH can assume more or less control over travel decisions. This flexibility allowed the development of multiple task-specific operating modes for the NavChair.

3.2 Vector Force Field (VFF)

A second obstacle avoidance routine intended for use in combination with MVFH is the Vector Force Field (VFF) method [9]. Like VFH, VFF was originally developed for round autonomous robots. The VFF method has been enhanced to work with irregularly shaped mobile robots [6] and has been applied to the NavChair system, as well (see Figure 4). In escence, VFF works by allowing every object detected by the NavChair's sonar sensors to exert a repulsive force on the NavChair's direction of travel, modifying its path of travel to avoid collisions. The repulsive force exerted by each object is proportional to its distance from the vehicle.

To account for the NavChair's rectangular shape, five different points on the chair are subject to the repulsive forces. The repulsive forces at each of these five points is summed and this total repulsive force is used to modify the NavChair's direction of travel.

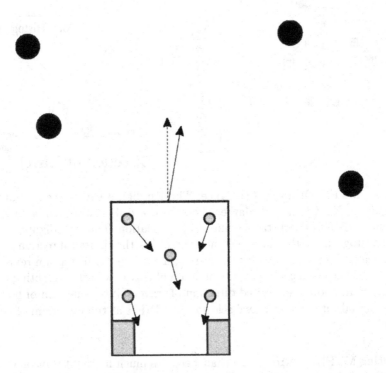

Fig. 4. Example of VFF Operating in The NavChair. The black circles represent obstacles, the gray circles are the five locations at which the repulsive forces are calculated, the lines extending from the gray circles represent the repulsive forces at each of these points (size of the arrows is proportional to magnitude of the repulsive force), the dashed line represents the direction the user pressed the joystick, and the solid line is the direction actually chosen by VFF.

4   Operating Modes

During the design of the NavChair system it became clear that in order to provide the desired range of functionality it would be necessary to define several different operating modes [10]. This section describes the function of each of the operating modes currently implemented within the NavChair. The results of several experiments are also presented to provide insight into the nature of each operating mode.

4.1   General Obstacle Avoidance (GOA) Mode

General Obstacle Avoidance (GOA) mode is the "default" operating mode of the NavChair. GOA mode is intended to allow the NavChair to quickly and smoothly navigate in crowded environments while maintaining a safe distance from obstacles. MVFH and VFF are both active in this mode. The weighting function used by MVFH is a relatively wide parabola (compared to the NavChair's other operating modes) centered on the joystick direction, which allows the chair a relatively large degree of control over the chair's direction of travel. This mode allocates the most control to the NavChair, in that it has great freedom in choosing a direction of travel to avoid obstacles while attempting to remain close to the direction indicated by the user.

A simple experiment was performed to analyze GOA mode's ability to successfully navigate the NavChair through a crowded room [11]. The experimental environment is shown in Figure 5. An able- bodied subject performed ten trials with the NavChair in GOA mode and ten trials with no navigation assistance active (in other words, the NavChair behaved exactly like a normal power wheelchair). In each trial the subject's task was to follow the path indicated in Figure 5. The results of the experiment are shown in Table 2.

Table 2. Results from Experiment Comparing General Obstacle Avoidance Mode with No Navigation Assistance

Measure	General Obstacle Avoidance	No Navigation Assistance
Average Time (sec)	9.35	7.09
Average Speed (mm/sec)	606.19	758.72
Average Minimum Obstacle Clearance (mm)	591.74	526.06

As can be seen from Table 2, GOA mode caused the NavChair to move more slowly through the slalom course than was possible when navigation assistance was not active. However, the NavChair also maintained a greater minimum distance from obstacles in GOA mode, due to the influence of the NavChair's collision avoidance routines. It is important to note that the NavChair assistive navigation system is designed to assist people who might not otherwise be able to operate a power wheelchair. Thus, while it may slow the wheelchair down for a "best case" able-bodied user, it can also provide a level of performance not otherwise achievable for users whose impairments limit their ability to operate a powered wheelchair.

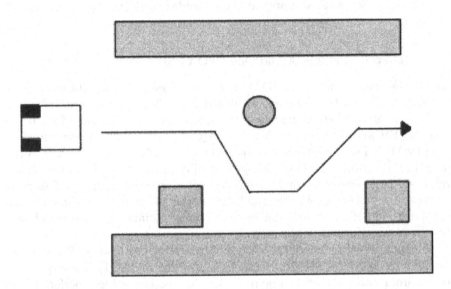

Fig. 5. General Obstacle Avoidance vs. No Navigation Assistance

4.2   Door Passage (DP) Mode

Door Passage (DP) mode is intended for use in situations requiring the NavChair to move between two closely spaced obstacles, such as the posts of a doorway. DP mode acts to center the NavChair within a doorway and then steer the chair through it. In this mode, VFF is not active and MVFH's weighting function is a narrow parabola, forcing the NavChair to adhere closely to the user's chosen direction of travel.

Figure 6 shows the operation of DP mode. As the chair passes through the doorway, MVFH acts to push the chair away from both doorposts and towards the center of the door. MVFH also acts to reduce the chair's speed as it approaches the doorway. If the user points the joystick in the general direction of a door, the effect is to funnel the NavChair to the center and through an open doorway.

Due to the influence of obstacle avoidance, it is possible for the NavChair to fail to successfully pass through a doorway on a given attempt. Typically, this is due to the NavChair approaching the door at an angle rather than from directly in front of the door. When a failure occurs the operator is then forced to back up and approach the door again, hopefully from a better direction.

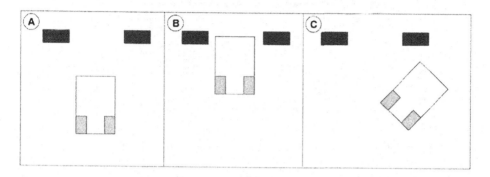

Fig. 6. Door Passage Mode. Panel A shows a situation that would prompt the NavChair to enter ADP mode. If the wheelchair operator directs the NavChair towards the door, ADP mode will act to center the chair in the doorway and move the chair through the door (Panel B). However, if the wheelchair operator directs the chair away from the door, ADP mode will not push the chair through the door (Panel C).

An experiment was performed to compare the ability of GOA mode and DP mode to pass between closely spaced obstacles [11]. In this experiment an able-bodied subject attempted to steer the NavChair through a door whose width was varied. Twenty trials were performed at each width. In ten of the trials the NavChair was in GOA mode and in ten of the trials the NavChair was in DP mode. The results of the experiment are shown in Figure 7.

As can be seen from the graph, DP mode allows the NavChair to pass through significantly smaller spaces than GOA. Of particular interest, the NavChair successfully passed through spaces 32 inches (81.3 cm) wide 70% of the time. This is noteworthy because the federal Architectural and Transportation Barriers Compliance Board (1984) has declared 32 inches as the minimally acceptable door width for wheelchair accessibility in federal buildings. With no navigation assistance active, the NavChair is able to pass through doorways as small as 25 inches (63.5 cm), which corresponds to the width of the NavChair. This corresponds with the "best case" scenario in which navigation assistance is neither needed by the user nor provided by the NavChair. Once again, the NavChair's navigation assistance ability does not (nor is it expected to) match the performance of an able-bodied user, but does provide sufficient navigation assistance to allow users with difficulty operating a standard power wheelchair to successfully perform tasks such as passing through doorways.

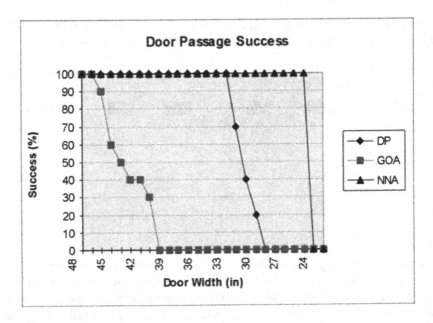

Fig. 7. Results From an Experiment Comparing the Performance of Door Passage Mode, General Obstacle Avoidance Mode, and No Navigation Assistance on a Door Passage Task. DP = Door Passage Mode, GOA = General Obstacle Avoidance Mode, NNA = No Navigation Assistance.

4.3   Automatic Wall Following (AWF) Mode

Automatic Wall Following (AWF) mode causes the NavChair to modify the user's joystick commands to follow the direction of a wall to the left or right of the chair. In this mode neither MVFH nor VFF is active. Instead, the NavChair uses the sonar sensors to the front and side opposite the wall being followed to scan for obstacles while the remaining sonar sensors (facing the wall) are used to navigate the chair. The NavChair's speed is reduced in proportion to the distance to the closest detected obstacle, which allows the NavChair to stop before a collision occurs.

Figure 8 shows the operation of AWF mode. As long as the user points the joystick in the approximate direction of the wall being followed, the chair modifies the direction of travel to follow the wall while maintaining a safe distance from the wall. However, if the user points the joystick in a direction sufficiently different from that of the wall then the user's direction is followed instead.

An experiment was performed to compare the performance of the NavChair operating in GOA mode, AWF mode, and without navigation assistance in a hallway traversal task [11]. In this experiment an able-bodied subject performed thirty trials in which he attempted to navigate the NavChair down an empty hallway. In ten of the trials the NavChair was in GOA mode and the subject

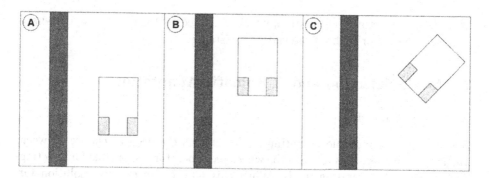

Fig. 8. Automatic Wall Following Mode. Panel A shows a situation which is appropriate for the NavChair to use AWF mode. If the user continues to direct the chair along a path roughly parallel to the wall, the NavChair will follow the direction of the wall (Panel B). However, if the user directs the chair in a direction sufficiently different from the wall, the NavChair will leave AWF mode and move away from the wall. The sonar sensors facing the wall are used to follow the wall while the sonar sensors in front of the chair are used to scan for obstacles.

moved the NavChair down the hallway by pointing the joystick at a 45 degree angle to the wall. In the second set of ten trials the NavChair was in AWF mode. In the final set of ten trials, the NavChair's navigation assistance was not active. The results of the experiment are shown in Table 3.

Table 3. Results of an Experiment Comparing the Performance of Automatic Wall Following Mode, General Obstacle Avoidance mode, and No Navigation Assistance on a Hallway Traversal Task

Measure	Automatic Wall Following	General Obstacle Avoidance	No Navigation Assistance
Average Time (sec)	9.13	11.27	4.6
Average Speed (mm/sec)	763.90	630.99	1447.17
Average Minimum Obstacle Clearance (mm)	407.38	556.56	322.25

As can be seen from the results of this experiment, AWF mode allows the NavChair to travel at a faster speed closer to a wall than GOA mode can but does not allow the chair to travel as fast or as close to the wall as is possible for an able-bodied operator using the chair without navigation assistance. However, AWF is expected to provide a measureable improvement in performance for the

NavChair's target user population, which is defined by their inability to operate a power wheelchair without navigation assistance.

5 Mode Selection and Automatic Adaptation

5.1 Introduction

The presence of multiple operating modes creates the need to choose between them. One alternative is to make the wheelchair operator responsible for selecting the appropriate operating mode. While this may be an effective solution for some users, it would present unreasonable demands for others. Alternatively, a method for the NavChair to automatically select the correct operating mode on its own has been developed [11]. This method combines information from two distinct adaptation methods. The first, Environmentally-Cued Adaptation (ECA), is based on information about the NavChair's immediate surroundings. The second, Location-Based Adaptation (LBA), is based on information from a topological map of the area in which the NavChair is located.

5.2 Combining ECA with LBA

Information from ECA and LBA is combined using a probabilistic reasoning technique known as Bayesian networks [12]. Bayesian networks use probabilistic information to model a situation in which causality is important, but our knowledge of what is actually going on is incomplete or uncertain. Bayesian networks can be thought of as a means of organizing information to allow the convenient application of a form of Bayes' theorem:

$$\Pr(H \mid e) = \frac{\Pr(e \mid H)\Pr(H)}{\Pr(e)}$$

where, in our applications, H represents the NavChair's operating modes, e is the set of observations, $\Pr(e \mid H)$ represents the probability of observing the most recent evidence given that a particular operating mode, and $\Pr(H \mid e)$ represents the probability that a particular operating mode is the most appropriate operating mode given the available evidence.

Because Bayesian network reasoning is based on probabilistic information, they are well-suited for dealing with exceptions and changes in belief due to new information. An additional advantage is that a network's architecture and internal values provide insight into the nature and connections of the information sources being used to derive conclusions. While none of this precludes the use of other methods, it does make Bayesian networks an attractive option.

Figure 9 shows the Bayesian Network which is used to combine LBA information with that from ECA . For computational efficiency, the Bayesian network is not explicitly represented within the NavChair. Instead, the Bayesian network is "reduced" to a series of parametric equations that receive evidence vectors as input and produce the belief vector for the Correct Operating Mode node as output. Part of the process of reducing the Bayesian network is converting the multiply-connected network in Figure 9 to the equivalent singularly-connected network in Figure 10.

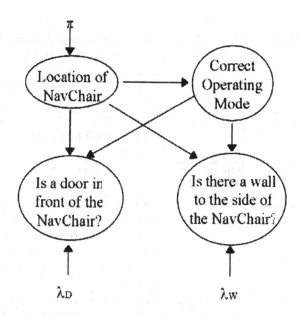

Fig. 9. Bayesian Network Used for Adaptation Decisions in The NavChair

To facilitate understanding, Table 4 contains explanations of all of the symbols used in the following explanation of the Bayesian network. The prior probability vector, π, contains the probability of being in each of the locations specified in the internal map. The two posterior evidence vectors, λ_D and λ_W, contain the probabilities of observing the current sonar signals (in other words, the observed evidence, e given that the environment contained either a door or a wall $(\Pr(e \mid Door = TRUE)$ and $\Pr(e \mid Wall = TRUE))$, respectively).

Table 4: Symbols Used in Explanation of Bayesian Network

Symbol	Type	Name	Explanation
π	vector	Prior Probability Vector	Contains the probability of being in each of the locations specified within the topological map.
λ_D, λ_W	vector	Posterior Evidence Vector	Contains the conditional probabilities of observing the most recent sonar readings given that there is a door/wall in front of the NavChair.

Table 4: (continued)

Symbol	Type	Name	Explanation
e	vector	Observed Evidence Vector	The most recent set of sonar readings.
$\pi(LM)$	vector	Prior Probability Vector	Contains the probabilities that the current location is l_a and the current operating mode is m_b for all combinations of locations $(l_1, \ldots, l_i)$ and operating modes $(m_1, \ldots, m_j)$.
l_a	scalar	Location	The ath of i $(1 < a < i)$ possible locations specified by the NavChair's topological map.
m_b	scalar	Operating Mode	The bth of j $(1 < b < j)$ total operating modes.
$M_{D|LM}$	matrix	Conditional Probability Matrix	The conditional probability matrix for the Door node (see Figure 10). Each element of the matrix represents the probability of the sonar sensors finding a door given a particular location (l_a) and operating mode (m_b).
$M_{W|LM}$	matrix	Conditional Probability Matrix	The conditional probability matrix for the Wall node (see Figure 10). Each element of the matrix represents the probability of the sonar sensors finding a wall given a particular location (l_a) and operating mode (m_b).
$M_{M|L}$	matrix	Conditonal Probability Matrix	The conditional probability matrix for the Correct Operating Mode node (see Figure 10). Each element of the matrix represents the probability of a particular operating mode (m_b) being the correct operating mode given that the NavChair is in a particular location (l_a).
$\pm d$	probabilistic variable	Door/No Door	A door is (not) observed by the sonar sensors.
$\pm w$	probabilistic variable	Wall/No Wall	A wall is (not) observed by the sonar sensors.
e^+	set	Prior Evidence	Evidence used to determine what location the NavChair is in.
e^-	set	Observed Evidence	Sonar sensor readings.

Table 4: (continued)

Symbol	Type	Name	Explanation
$\lambda_D(LM)$	vector	Evidence Vector	Contains the probability of observing the most recent sonar evidence pertaining to the presence of a door in front of the NavChair given all combinations of locations $(l_1, \ldots, l_i)$ and operating modes $(m_1, \ldots, m_j)$.
$\lambda_W(LM)$	vector	Evidence Vector	Contains the probability of observing the most recent sonar evidence pertaining to the presence of a wall to the side of the NavChair given all combinations of locations $(l_1, \ldots, l_i)$ and operating modes $(m_1, \ldots, m_j)$.
$BEL(LM)$	vector	Belief Vector	Contains the probability that the location is la and the correct operating mode is mb given all combinations of locations $(l_1, \ldots, l_i)$ and operating modes $(m_1, \ldots, m_j)$.
$BEL(M)$	vector	Belief Vector	Contains the probability that the correct operating mode is mb for all operating modes $(m_1, \ldots, m_j)$.
L	set	Set of All Locations	Contains all locations $(l_1, \ldots, l_i)$ specified by the topological map.

Every time the NavChair makes an adaptation decision, the location of the NavChair in the internal map is used to construct the π vector, and the output of the processes for identifying doorways and walls from the NavChair's sonar sensors are used to create the λ_D and λ_W vectors.

Evaluating the network in Figure 10 requires the specification of three conditional probability matrices, one for each node. The conditional matrix for the Door node takes the form:

$$
M_{D|LM} = \begin{bmatrix} \Pr(+d \mid l_1 m_1) & \Pr(-d \mid l_1 m_1) \\ \vdots & \vdots \\ \Pr(+d \mid l_i m_j) & \Pr(-d \mid l_i m_j) \end{bmatrix} \tag{1}
$$

where $\Pr(+d \mid l_1 m_1)$ represents the probability of observing a door $(+d)$ given that the NavChair is in location 1 (l_1) out of i possible locations and the correct operating mode is m_1 out of j possible operating modes.

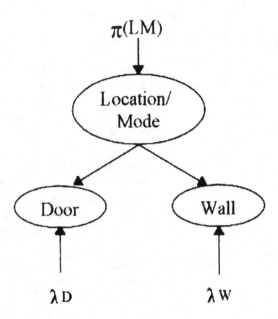

Fig. 10. Equivalent Bayesian Network Used for Mode Decisions

The conditional matrix for the Wall node is of the form:

$$M_{D|LM} = \begin{bmatrix} \Pr(+w \mid l_1 m_1) \ \Pr(-w \mid l_1 m_1) \\ \vdots \qquad\qquad \vdots \\ \Pr(+w \mid l_i m_j) \ \Pr(-w \mid l_i m_j) \end{bmatrix} \tag{2}$$

and can be interpreted similarly to the conditional probability matrix for the Door operating node.

Finally, the conditional matrix for the Mode node which is combined with the Location node in Figure 10 is also needed:

$$M_{M|L} = \begin{bmatrix} \Pr(m_1 \mid l_1) \ \cdots \ \Pr(m_j \mid l_1) \\ \vdots \qquad \ddots \qquad \vdots \\ \Pr(m_1 \mid l_i) \ \cdots \ \Pr(m_j \mid l_i) \end{bmatrix} \tag{3}$$

where $\Pr(m_1 \mid l_1)$ represents the probability that m_1 is the correct operating mode given that the NavChair is in location l_1, again out of j possible modes and i possible locations.

The process of making a mode decision in the NavChair proceeds as follows:

1. The system updates the contents of λ_D and λ_W based on the probability of obtaining the most recent sonar data if a door was in front of the chair, $\Pr(e \mid D)$, and the probability of obtaining the most recent sonar data if a wall were to the right or left of the chair, $\Pr(e \mid W)$.

$$\lambda_D = \begin{bmatrix} \Pr(e^- \mid +d) \\ \Pr(e^- \mid -d) \end{bmatrix} \tag{4}$$

and

$$\lambda_D = \begin{bmatrix} \Pr(e^- \mid +w) \\ \Pr(e^- \mid -w) \end{bmatrix} \tag{5}$$

2. The system updates the contents of π based on the location of the chair. If the chair is in location k within the map,

$$\pi = \left[\Pr(l_1 \mid e^+) \cdots \Pr(l_1 \mid e^+) \right]^T$$
$$= [0 \cdots 010 \cdots 0]^T \tag{6}$$

where the kth element of π is 1.

3. The effects of the observed evidence are propogated upwards towards the Location/Mode node. The vector from the Door node is calculated by:

$$\lambda_D(LM) = M_{D|LM} \bullet \lambda_D$$
$$= \begin{bmatrix} \Pr(e_D^- \mid l_1 m_1) \\ \vdots \\ \Pr(e_D^- \mid l_i m_j) \end{bmatrix} \tag{7}$$

Similarly, the vector from the Wall node is calculated to be

$$\lambda_W(LM) = M_{W|LM} \bullet \lambda_W$$
$$= \begin{bmatrix} \Pr(e_W^- \mid l_1 m_1) \\ \vdots \\ \Pr(e_W^- \mid l_i m_j) \end{bmatrix} \tag{8}$$

4. The effects of the prior evidence are propogated downward to the Location/Mode node.

$$\pi(LM) = \pi \bullet M_{M|L}$$
$$= \begin{bmatrix} \Pr(l_1 m_1 \mid e^+) \\ \vdots \\ \Pr(l_i m_j \mid e^+) \end{bmatrix} \tag{9}$$

5. The belief vector for the Location/Mode node is calculated based on the prior and observed evidence by the following formula:

$$
\begin{aligned}
BEL(LM) &= \alpha \bullet \pi(LM) \bullet \lambda(LM) \\
&= \alpha \bullet \pi(LM) \bullet [\lambda_D(LM) \bullet \lambda_W(LM)] \\
&= \begin{bmatrix} \Pr(l_1 m_1 \mid e) \\ \vdots \\ \Pr(l_i m_j \mid e) \end{bmatrix}
\end{aligned} \tag{10}
$$

6. The belief in each mode is then calculated from the Mode/Location belief vector:

$$
BEL(M) = \begin{bmatrix} \sum_{L=1}^{i} \Pr(lm_1 \mid e) \\ \vdots \\ \sum_{L=1}^{i} \Pr(lm_j \mid e) \end{bmatrix} \tag{11}
$$

7. The NavChair's operating mode is then chosen based on which element of BEL(M) has the highest value.

The final detail to be discussed is the selection of values for the conditional probability matrices. These values are filled in beforehand based on the environment in which the NavChair is operating and the nature of the task that it is expected to perform. When the NavChair moves between different environments, or the task(s) it is expected to accomplish changes, then the values of these matrices must be changed as well. There is currently no mechanism for the NavChair to automatically determine the values for these matrices.

5.3   Empirical Evaluation

The NavChair's automatic adaptation mechanism must meet several design criteria [13], the most important being that the method must make the correct operating mode decision as often as possible. In two experiments [11], the NavChair's automatic adaptation mechanism (ECA+LBA) performed better than ECA alone and compared favorably to an expert human making adaptation decisions.

Another important criterion is that the NavChair avoid frequent mode changes, which could lead to an uncomfortable ride for the operator. The NavChair's adaptation mechanism contains built in controls that limit the frequency with which it can change modes, which limit the possibility that it will rapidly switch between different operating modes.

Decisions must also be made in real-time. When in use, the NavChair's automatic adaptation mechanism does not interfere with normal operation of the wheelchair. In particular, the low number of collisions experienced during experiments implies that the NavChair was able to devote most of its computational resources to providing navigation assistance to the operator rather than making adaptation decisions.

6   Discussion

The NavChair has yet to be formally evaluated in trials involving individuals from its target user population. However, feedback has been sought from clinicians active in wheelchair seating and mobility during all phases of the NavChair's design, and an informal session with a potential user provided encouraging results. In our experience, when a standard wheelchair joystick is used to control the NavChair, the effects of the NavChair's navigation assistance tends to improve the performance of individuals that have difficulty operating a power wheelchair but tends to hinder the performance of individuals that do not have difficulty operating a power wheelchair. The primary reason that navigation assistance interferes with skilled driving performance is the tendency for navigation assistance to reduce the wheelchair's speed. Another problem arises from the lack of resolution provided by the NavChair's sonar sensors. A skilled wheelchair operator, guided by visual feedback can steer much closer to obstacles without fear of collision than is possible for the NavChair's software guided by sonar sensors. This results in the NavChair maintaining a much greater minimum distance from obstacles than is strictly necessary.

Future work is planned in several areas. First, there is a need to add additional operating modes to the NavChair. A close approach mode is already envisioned which will allow a user to "dock" the NavChair at a desk or table. The NavChair is also an attractive testbed for exploring alternative wheelchair interfaces. The NavChair can be used to examine the effects of different input (voice) and feedback (auditory and visual) options that are currently unavailable on standard power wheelchairs.

There still remains much work to be done on the NavChair's automatic adaptation mechanism. In particular, additional information sources need to be identified and incorporated into the existing Bayesian network. One likely information source is user modeling. Some work [9] has already been performed in this area which must be expanded upon before it can be included in the Bayesian network. There is a need to add additional operating modes to the NavChair. A close approach mode is already envisioned which will allow a user to "dock" the NavChair at a desk or table.

There is also a need to add more environmental sensors to the NavChair. Currently, the NavChair has very few sensors on its sides and does not have any sensors at all on its back. This can cause the NavChair to become confused when moving within a tightly confined area. In addition to sonar sensors, infrared range finders and bump sensors should be added to the NavChair to improve the capability of its obstacle avoidance routines.

Finally, there is a need for formal testing of the NavChair with individuals with disabilities. This will require that the NavChair be modified to accommodate the multitude of seating and positioning hardware that members of its target user population normally employ. In addition, the NavChair will also have to accommodate a larger variety of input methods, such as head joysticks, pneumatic controllers, and switch arrays.

References

1. S. Levine, Y. Koren, and J. Borenstein. The navchair control system for automatic assistive wheelchair navigation. In *Proceedings of the 13th Annual RESNA Conference*, pages 193–194, Washington, D.C., 1990. RESNA.

2. L. Jaros, D. Bell, S. Levine, J. Borenstein, and Y. Koren. Navchair: Design of an assistive navigation system for wheelchairs. In *Proceedings of the 17th Annual RESNA Conference*, pages 379–381, Las Vegas, NV, 1993. RESNA.

3. J. Borenstein and Y. Koren. Noise rejection for ultrasonic sensors in mobile robot applications. In *Proceedings of the IEEE Conference on Robotics and Automation*, pages 1727–1732, New York, 1992. IEEE Press.

4. J. Borenstein and Y. Koren. Histogramic in-motion mapping for mobile robot obstacle avoidance. *IEEE Transactions on Robotics and Automation*, 17(4):535–539, 1991.

5. J. Borenstein and Y. Koren. The vector field histogram - fast obstacle avoidance for mobile robots. *IEEE Journal of Robotics and Automation*, 7(3):278–288, 1989.

6. J. Borenstein and U. Raschke. Real-time obstacle avoidance for non-point mobile robots. In *Proceedings of the Fourth World Conference on Robotics Research*, pages 2.1–2.9, Pittsburgh, PA., 1991.

7. D. Bell, S. Levine, Y. Koren, L. Jaros, and J. Borenstein. An assistive navigation system for wheelchairs based upon mobile robot obstacle avoidance. In *Proceedings of the IEEE International Conference on Robotics and Automation*, pages 2018–2022, New York, 1994. IEEE Press.

8. D. Bell, S. Levine, Y. Koren, L. Jaros, and J. Borenstein. Design criteria for obstacle avoidance in a shared-control system. In *Proceedings of the RESNA International Conference*, pages 581–583, Washington, D.C., 1994. RESNA.

9. D. Bell. *Modeling Human Behavior For Adaptation in Human Machine Systems*. PhD thesis, University of Michigan, 1994.

10. D. Bell, S. Levine, Y. Koren, L. Jaros, and J. Borenstein. Shared control of the navchair obstacle avoiding wheelchair. In *Proceedings of the RESNA International Conference*, pages 370–372, Washington, D.C., 1993. RESNA.

11. R. Simpson. *Improved Automatic Adaptation Through the Combination of Multiple Information Sources*. PhD thesis, University of Michigan, 1997.

12. E. Charniak. Bayesian networks without tears. *AI Magazine*, 12(4):50–63, 1991.

13. R. Simpson, S. Levine, and H. Koester. Using probabilistic reasoning to develop automatically adapting assistive technology systems. In *Proceedings of the AAAI Fall Symposium on Developing AI Applications for People with Disabilities*, pages 104–108, Boston, MA, 1996. AAAI.

Wheelesley: A Robotic Wheelchair System: Indoor Navigation and User Interface

Holly A. Yanco

MIT Artificial Intelligence Laboratory
545 Technology Square, Room 705
Cambridge, MA 02139
U.S.A.
holly@ai.mit.edu

Abstract. Many people in wheelchairs are unable to control a powered wheelchair with the standard joystick interface. A robotic wheelchair can provide users with driving assistance, taking over low-level navigation to allow its user to travel efficiently and with greater ease. Our robotic wheelchair system, Wheelesley, consists of a standard powered wheelchair with an on-board computer, sensors and a graphical user interface. This paper describes the indoor navigation system and the customizable user interface.

1   Introduction

The goal of the Wheelesley project is the development of a robotic wheelchair system that provides navigational assistance in indoor and outdoor environments, which allows its user to drive more easily and efficiently. A robotic wheelchair is usually a semi-autonomous system, which means that a full solution to Artificial Intelligence problems do not need to be found before a useful system can be built. A robotic wheelchair can take advantage of the intelligence of the chair's user by asking for help when the system has difficulty navigating.

There are two basic requirements for any robotic wheelchair system. First and foremost, a robotic wheelchair must navigate safely for long periods of time. Any failures must be graceful to prevent harm from coming to the user. Second, in order for such a system to be useful, it must interact effectively with the user. Outside of these two requirements, desirable features may include outdoor as well as indoor navigation, automatic mode selection based upon the current environment and task to reduce the cognitive overhead of the user, and easily adaptable user interfaces.

The Wheelesley system takes over low-level navigation control for the user, allowing the user to give higher level directional commands such as "forward" or "right." Most people take low-level control for granted when walking or driving. For example, when walking down a busy corridor, a person is not usually aware of all of the small changes he makes to avoid people and other obstacles. However,

V. O. Mittal et al. (Eds.): Assistive Technology and AI, LNAI 1458, pp. 256–268, 1998.
© Springer-Verlag Berlin Heidelberg 1998

for users in our target community, low-level control requires just as much effort as high-level control. For example, it may be easy for a disabled person to gesture in the direction of a doorway, but it may be difficult for that person to do the fine navigation required to direct the wheelchair through a doorway that is barely wider than the wheelchair. The robot carries out each command from the user by using its sensors and control code to safely navigate.

The Wheelesley robotic wheelchair system is a general purpose navigational assistant in environments that are accessible for the disabled (e.g., ramps and doorways of sufficient width to allow a wheelchair to pass). The reactive system does not use maps for navigation. One of the advantages of this strategy is that users are not limited to one particular location by the need for maps or environment modifications. This paper describes indoor navigation in the Wheelesley system; outdoor navigation is currently under development.

The target community for this system consists of people who are unable to drive a powered wheelchair by using a standard joystick. The users vary in ability and access methods. Some people can move a joystick, but are unable to make fine corrections to movement using the joystick. Other people are able to click one or more switches using their head or other body part. Some of our potential users are unable to control a powered wheelchair with any of the available access devices. The wide variety of user abilities in our target community requires that the system be adaptable for many types of access devices.

While members of the target community have different abilities, we assume that all users will have some common qualities. We expect that any potential user can give high-level commands to the wheelchair through some access method and a customized user interface. We assume that the user of the wheelchair is able to see, although later versions of the system may be developed for the visually impaired. We also assume that a potential user has the cognitive ability to learn to how to operate the system and to continue to successfully operate the system once out of a training environment.

2   Related Work

This research differs from previous research in robotic wheelchairs and mobile robots in four ways. (Some systems have incorporated some of these issues, but none has incorporated all of them.) First, it will be able to navigate in indoor and outdoor environments, switching automatically between the control mode for indoor navigation and the control mode for outdoor navigation. Second, it is a reactive system and does not require maps or planning. The system can be used in a variety of locations, giving the user more freedom. Third, interaction between the user and the wheelchair is investigated. The robot should provide feedback to the user as it makes navigation decisions and should ask for additional information when it is needed. Finally, the system has an easily customizable user interface. A wide range of access methods can be used to control the system.

Over the years, several robotic wheelchair systems have been developed (see [5] for an overview of assistive robotics). Some of the previous research on robotic wheelchairs has resulted in systems that are restricted to a particular location. One example of restrictive assistive wheelchairs are systems that rely on map-based navigation. The system will perform efficiently only when a complete and accurate map is available; the map can either be provided to or created by the robot. The system will either fail to work or work inefficiently when the robot is operating in an environment for which it has no map. If the robot can only operate efficiently in one building (as, e.g., [8]), the user will not be able to use the technology once he leaves the doorway of the known building. Since most people need to be in several buildings during one day, this system is not general enough, although it is a step towards assistive robotics. Even more restrictive than a map-based system is that of [10]. This system requires the use of a magnetic ferrite marker lane for navigation. Once the wheelchair's user leaves the magnetic path, the technology of the assistive system is useless.

The NavChair [9] does not restrict its user to a particular building, but it does restrict its user to an indoor environment. The NavChair navigates in indoor office environments using a ring of sonar sensors mounted on the wheelchair tray. The height of the sensors prevents the system from being used outdoors since it can not detect curbs. People who are unable to drive a standard powered wheelchair have been able to drive the NavChair using sensor guidance and either the joystick or voice commands.

A deictic navigation system has been developed to drive a robotic wheelchair [1]. This system navigates relative to landmarks using a vision-based system. The user of the wheelchair tells the robot where to go by clicking on a landmark in the screen image from the robot's camera and by setting parameters in a computer window. The robot then extracts the region around the mouse click to determine to which landmark the user wishes to travel. It then uses the parameters to plan and execute the route to the landmark. Deictic navigation can be very useful for a disabled person, but a complicated menu might be difficult to control with many of the standard access methods.

The TAO project [3] has developed a robotic module for navigation that can be interfaced with standard wheelchairs. The navigation module has been put on two different commercially available wheelchairs. The system uses computer vision to navigate in its environment. It is primarily an indoor system, although it has been tested outdoors in limited situations. The TAO wheelchairs navigate in an autonomous mode, randomly wandering in an environment. The user can override the robotic control by touching the joystick. In joystick mode, no assistance is provided.

3   Robot Hardware

The robotic wheelchair used in this research (Figure 1) was built by the KISS Institute for Practical Robotics [6]. The base is a Vector Mobility powered wheelchair. The drive wheels are centered on either side of the base, allowing the

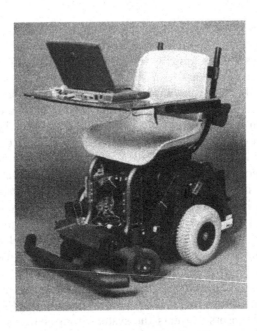

Fig. 1. Wheelesley, the robotic wheelchair system.

chair to turn in place. There are two front casters and a rear caster with spring suspension.

The robot has a 68332 processor that is used to control the robot and process sensor information. For sensing the environment, the robot has 12 SUNX proximity sensors (infrared), 4 ultrasonic range sensors, 2 shaft (wheel) encoders and 2 Hall effect sensors. The infrared and sonar sensors are placed around the perimeter of the wheelchair, with most pointing towards the front half of the chair (see Figure 2 for a map of the sensor placement). The Hall effect sensors are mounted on the front bumper of the wheelchair. Additional sensors to determine the current state of the environment (indoor or outdoor) are being added to the system.

A Macintosh Powerbook is used for the robot's graphical user interface. The focus was on creating an interface that could be easily customized for various users and their access methods (as described in Section 5).

4   A Navigation System for Indoor Environments

The focus of mobile robotics research is the development of autonomous navigation systems. However, a robotic wheelchair must interact with its user, making the robotic system semi-autonomous rather than completely autonomous. An autonomous mobile robot is often only given its goal destination and a map. A robotic wheelchair should not subscribe to this method. The user may decide to change course during traversal of the path – as he starts to go by the library on the way to the mail room, he may decide to stop at the library to look for a book. The wheelchair robot must be able to accept input from its user not only at the start of the trip, but throughout the journey. The robot should have the ability to take on a greater autonomous role if the user desires it, but the robot will still need to work in conjunction with the user. The user interface developed for this purpose is described below in Section 5.

The system uses reactive navigation becasue the user must be able to successfully navigate novel environments immediately. Because there is an intelligent human giving high-level navigation commands to the wheelchair robot, the common limitation of a reactive navigation system (lack of planning) is alleviated. The system concentrates on what a reactive system can do well by carrying out the user's commands while keeping the user safe, leaving the planning that a reactive system typically omits to the user. If interviews with members of the target community indicate that they want the robot to be more autonomous, maps of commonly traveled environments such as the home and the office could be incorporated. Path planning for indoor robotics has been studied extensively (see [4] for examples) and could be implemented on the robotic wheelchair base.

There are two types of control when driving a wheelchair: low-level and high-level. Low-level control involves avoiding obstacles and keeping the chair centered in a hallway. High-level control involves directing the wheelchair to a desired location. For a power wheelchair user who has good control of the joystick, these two types of control can be easily managed at the same time. The user can avoid obstacles on the path by moving the joystick to make the proper adjustment. This is analogous to driving a car; people make many small adjustments to their route to keep the car in the proper lane and to avoid obstacles like potholes.

When a power wheelchair user does not have perfect control of a joystick or has no control of a joystick at all, low-level control does not easily blend into high-level control. It is not possible to make small adjustments easily. For a user driving using an alternative access method (see Section 5.1), low-level control adjustments require as much effort as high-level directional commands. A robotic wheelchair can assist a user in this group by taking over low-level control, requiring the user to use the access method only to give high-level directional commands like "right" or "left."

In the Wheelesley system, the user gives the high-level commands ("forward," "left," "right," "back," and "stop") through the graphical user interface (see Section 5). The system carries out the user's command using common sense constraints such as obstacle avoidance. The robot's low-level control acts to keep the wheelchair and its user safe by using sensor readings. For example, if the user instructs the chair to go forward, the robot will carry out the command by taking over control until another command is issued. While executing the high-level "forward" command, the chair will prevent the user from running into walls or other obstacles. If the chair is completely blocked in front, it will stop and wait for another command from the user. If it is drifting to the right, it will correct itself and move to the left. This navigation method allows people who have trouble with fine motor control but who have the ability to issue high-level commands to control a powered wheelchair.

Indoor navigation relies on the infrared sensors, sonar sensors and Hall effect sensors. The infrared sensors give binary readings that indicate if something is about one foot from the sensor. As soon as an infrared sensor signals that it is blocked, the robot immediately corrects to avoid the obstacle. These close reading sensors function to avoid obstacles not anticipated by the sonar sensors. The sonar sensors return distance information. The sonar readings are smoothed over a short window of readings to diminish the effect of any noisy readings; the smoothed value is used to determine if there are obstacles too close to the wheelchair. The Hall effect sensors are mounted on the wheelchair's bumper and are used as a last resort. If an obstacle was missed by the infrared and sonar sensors while traveling forward, the bumper will hit the obstacle. Empirically, bumper hits are very infrequent (only one bumper hit in over ten hours of user testing).

The robot is able to traverse long hallways without requiring user corrections. The system uses infrared and sonar sensors pointed out to each side at a angle perpendicular to the forward movement. The system stays centered in the hallway by keeping sensor readings on each side of the chair equal. While moving down the hallway in this manner, the chair will also avoid obstacles in the path. Obstacle avoidance takes priority over the hallway centering. In designing the system, centering was chosen over wall following to keep the chair in a better position for turning to avoid obstacles.

5 Graphical User Interface

A robotic wheelchair system must be more than a navigation system. While it is important to develop a system that will keep its user from harm and assist in navigation, the system will be useless if it can not be adapted for its intended users. The Wheelesley system solves the adaptation problem through the addition of a general user interface that can be customized for each user.

The graphical user interface is built on a Macintosh Powerbook and can be easily customized for various access methods (see Section 5.1 for a discussion of access methods). To date, the interface has been customized for two access methods. The first is an eye tracking device called EagleEyes [2] (Section 5.2). The second is a single switch scanning device (Section 5.3).

The user interface is shown in Figure 2. (See [13] for an earlier version of the interface.) There are three control modes that the user can select. In manual mode, the joystick commands are passed directly to the motor controller with no sensor mediation. In joystick mode, the user's joystick commands are carried out using low-level control on the robot to avoid obstacles. In interface mode, the arrows are used to direct the robot. The navigation command portion of the interface used in interface control consists of four directional arrows and a stop button.

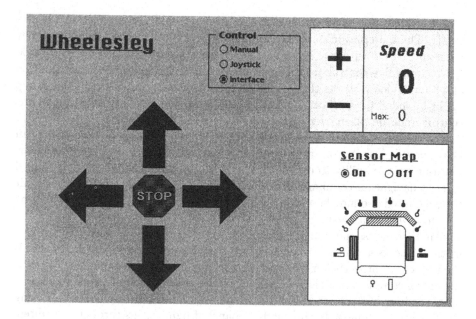

Fig. 2. The original user interface screen.

The user controls the standard speed of the robot by clicking on the plus and minus buttons in the upper right corner of the screen. The robot may move at a slower pace than the user requests when the current task requires a slower speed to be carried out safely. The actual speed of the robot is displayed by the robot under the speed control buttons.

The sensor map shows a representation of the wheelchair and the location of the sensors. Obstacles detected by the sensors are displayed on this sensor map. This is intended to provide a user who is unable to move his head with a picture of the obstacles in the world around him. In Figure 2, the sensor map is shown in the lower right corner. The rectangular bars represent the sonar sensors. The bar fills proportionally to indicate the distance of obstacles, filling more as obstacles get closer. An empty rectangle indicates that no object has been detected. The infrared sensors are represented by circles with a line pointed out in the sensor's direction of detection. For these binary detectors, an empty circle indicates that no obstacle has been detected and a full circle indicates that an obstacle has been detected.

This interface has been customized for two different access methods: eye tracking (see Section 5.2) and single switch scanning (see Section 5.3).

5.1   Access Methods

In the rehabilitation community, access methods are devices used to enable people to drive wheelchairs or control computers. Many different access methods for powered wheelchairs are currently used. The default access method is a joystick.

If a user has sufficient control with a joystick, no additional assistance is necessary. These users would not be candidates for a robotic wheelchair since they are able to drive without the system. If a person has some control of a joystick, but not very fine control, joystick movement can be limited through the addition of a plate which restricts the joystick to primary directions. Users in this group might be aided by a robotic system. If they push the joystick forward, the fine control could be taken over by the robotic system.

If a user is unable to use a joystick, there are other access devices which can be employed. A switch or group of switches can be used to control the wheelchair. If a user has the ability to use multiple switches, different switches can be linked to each navigation command. The multiple switches can be on the wheelchair tray, mounted around the user's head or placed anywhere that the user will be able to reliably hit them.

Another access method for wheelchairs is a sip and puff system. With this method, the user controls the wheelchair with blowing or sucking on a tube. If the user can control the air well enough, soft and hard sips or puffs can be linked to control commands. This is analagous to the multiple switch system above.

If the user has only one switch site, the wheelchair must be controlled using single switch scanning. In this mode, a panel of lights scans through four directional commands (forward, left, right and back). The user clicks the switch when the desired command is lit. If the user is traveling forward and drifts left, he must stop, turn the chair to the right and then select forward again. This mode of driving is very slow and difficult; it is the method of last resort. Obviously, a robotic wheelchair system could help this group of users.

Most research on robotic wheelchairs has not focused on the issue of access methods. Most of the current systems are driven using a joystick (e.g., [6], [3], and [9]). A few researchers have used voice control for driving a robotic wheelchair (e.g., [9]). Voice control can be problematic because a failure to recognize a voice command could cause the user to be unable to travel safely. Additionally, some members of our target community are non-verbal.

5.2   Customizing the User Interface for EagleEyes

Eye tracking has been investigated as a novel method for controlling a wheelchair. EagleEyes [2] is a technology that allows a person to control a computer through five electrodes placed on the head. Electrodes are placed above and below an eye and to the left and right of the eyes. A fifth electrode is placed on the user's forehead or ear to serve as a ground. The electrodes measure the EOG (electro-oculographic potential), which corresponds to the angle of the eyes in the head. The leads from these electrodes are connected to two differential electrophysiological amplifiers. The amplifier outputs are connected to a signal acquisition system for the Macintosh.

Custom software interprets the two signals and translates them into cursor coordinates on the computer screen. The difference between the voltages of the electrodes above and below the eye is used to control the vertical position of the cursor. The voltage difference of the electrodes to the left and right of the eyes controls the horizontal position of the cursor. If the user holds the cursor in a small region for a short period of time, the software issues a mouse click.

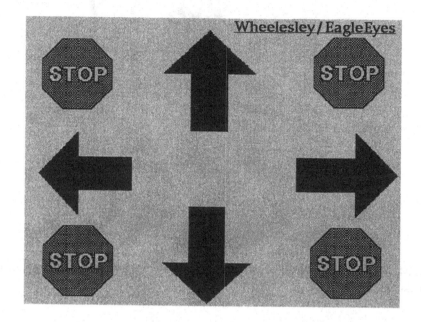

Fig. 3. The customized interface for use with EagleEyes.

The user interface was quickly customized for use with EagleEyes [11]. The screen (Figure 3) was redesigned to accommodate the needs of the EagleEyes system. Large buttons are easier to use with an electrode system than small ones. The interface has four large direction arrows and four large stop buttons. Four stop buttons are provided so that the user will be near a stop button regardless of the current cursor position. To move, the user moves the cursor to the appropriate arrow through eye and head movement and dwells on the arrow to issue a mouse click. The robot travels in the commanded direction, avoiding obstacles and staying centered in the hallway, until a new directional command or a stop command is issued.

The robotic wheelchair has been successfully controlled by three able-bodied subjects using EagleEyes. (See Figure 4 for a photo of the two systems being used together.)

There is currently a "Midas Touch"-like problem with this access method; there is no way for the computer to differentiate between the cursor moving because the user wants to issue a command and the cursor moving because the user is looking around the environment. The able-bodied subjects solved this problem by fixing their gaze either on the arrow for the current direction or on part of the unused portion of the screen. Other users may not be as proficient with EagleEyes and might look at other command buttons accidentally. This

Fig. 4. The robotic wheelchair system being driven using EagleEyes, an eye tracking system.

problem could be solved by using a voluntary blink as a mouse click or by using a voluntary blink to switch in and out of using EagleEyes to control the wheelchair. Another solution would be to use a single switch for users able to reliably hit a switch; the switch could be used to toggle the modes or could be used as a mouse click.

With an eye tracker as a control method, an experienced user may not need to have a computer screen in front of him on a tray. Once the user learned how to issue commands on the screen, the user could move his head and eyes in a similar manner to issue commands with the screen removed. This would make the robotic wheelchair look more like a standard wheelchair, which is desired by many potential users.

5.3 Customizing the User Interface for Single Switch Scanning

Single switch scanning is the access method of last resort for traditional powered wheelchairs. A single switch scanning system consists of a switch and a control panel with four lights for four directions (forward, left, right and back). When using this method, the control panel scans through the four commands. The user clicks the single switch when the control panel shows the desired direction. Usually, these systems are not "latched" for forward. This means that person must hold down the switch as long as he wishes to go forward. Latching the system would mean the wheelchair would start going forward when the switch was pressed and would continue going forward until the switch is pressed again.

This is considered too dangerous for a standard powered wheelchair configuration since the wheelchair would continue to drive if the user was unable to press the switch to stop it.

This method is very difficult to use for traditional powered wheelchairs, primarily because drift is a significant problem. To correct a drift to the left of the right, the user must stop going forward, wait for the scanning device to get to the arrow for the direction of choice, click to turn the chair, stop turning, wait to scan to forward and then click to move forward again. Robotic assisted control can improve driving under this access method by correcting drift automatically and avoiding obstacles. Additionally, the system can be latched due to the safety provided by robotic control.

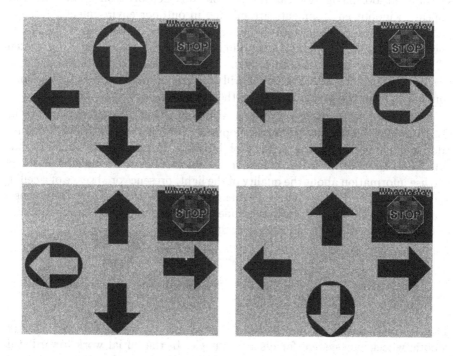

Fig. 5. The customized interface for single switch scanning. The interface scans through the four directions in the following order: forward, right, left and back. To start moving, the user clicks a switch when the interface is highlighting the arrow corresponding to the desired direction.

Customization for this access method took less than 1 hour. The screen has four arrows and one stop button (see Figure 5). The system scans in the same pattern as commercially available single switch scanning systems (forward, right, left, back). The stop button is only on the screen so that it can be highlighted when the chair is stopped. User tests with fifteen able-bodied subjects determined that an obstacle course can be completed in less time and with less effort with navigational assistance than without. Users traversed the obstacle course in 25% less time with 71% fewer clicks with robotic assisted control. (See [12] for details.)

6 Future Work

Work is continuing towards the goal of a complete robotic wheelchair system. A robotic wheelchair must be able to navigate in both indoor and outdoor environments. While indoor navigation can be successful with infrared and sonar sensors, outdoor navigation can not rely on these sensors alone. The walls that appear in indoor environments are missing in outdoor environments. A vision system for outdoor navigation is being developed. The philosophy of taking over low-level control is also being followed in the outdoor domain. The vision system will locate sidewalks, curbs, curb cuts, crosswalks, handicap ramps and obstacles on the current path. The robot will continue to take high-level directional commands from the user and execute them while keeping the user safe.

The system will automatically select indoor or outdoor mode using an indoor/outdoor sensor currently in development. For a user unable to use a mouse, adding extra items to the screen is prohibitive. A user can not be asked to indicate when he has traveled from indoors to outdoors or the reverse. The sensor will use information about the quality of the light, presence or absence of a ceiling and temperature data to determine the current state of the robot. This sensor could also be extended to select submodes in an outdoor or indoor environment to optimize the selection of navigation code.

7 Summary

This research project is aimed towards developed a usable, low-cost assistive robotic wheelchair system for disabled people. In the initial work towards this goal, an indoor navigation system and a graphical user interface have been developed. The robotic wheelchair must work with the user to accomplish the user's goals, accepting input as the task progresses, while preventing damage to the user and the robot.

Acknowledgments

This research is funded by the Office of Naval Research under contract number N00014-95-1-0600, the National Science Foundation under grant number CDA-9505200 and a faculty research grant from Wellesley College.

References

1. J.D. Crisman and M.E. Cleary. Progress on the deictically controlled wheelchair. In Mittal et al. [7]. This volume.
2. J. Gips. On building intelligence into eagleeyes. In Mittal et al. [7]. This volume.
3. T. Gomi and A. Griffith. Developing intelligent wheelchairs for the handicapped. In Mittal et al. [7]. This volume.
4. D. Kortenkamp, R.P. Bonasso, and R. Murphy, eds. *Artificial intelligence and mobile robots: case studies of successful robot systems*. MIT/AAAI Press, 1998.
5. D.P. Miller. Assistive robotics: an overview. In Mittal et al. [7]. This volume.
6. D.P. Miller and M.G. Slack. Design and testing of a low-cost robotic wheelchair prototype. *Autonomous Robots*, 2:77–88, 1995.
7. V. Mittal, H.A. Yanco, J. Aronis and R. Simpson, eds. *Lecture Notes in Artificial Intelligence: Assistive Technology and Artificial Intelligence*. Springer-Verlag, 1998. This volume.
8. M.A. Perkowski and K. Stanton. Robotics for the handicapped. In *Northcon Conference Record*, pages 278–284, 1991.
9. R.C. Simpson, S.P. Levine, D.A. Bell, L.A. Jaros, Y. Koren and J. Borenstein. Navchair: an assistive wheelchair navigation system with automatic adaptation. In Mittal et al. [7]. This volume.
10. H. Wakaumi, K. Nakamura and T. Matsumura. Development of an automated wheelchair guided by a magnetic ferrite marker lane. *Journal of Rehabilitation Research and Development*, 29(1):27–34, Winter 1992.
11. H.A. Yanco and J. Gips. Preliminary investigation of a semi-autonomous robotic wheelchair directed through electrodes. In S. Sprigle, ed, *Proceedings of the Rehabilitation Engineering Society of North America 1997 Annual Conference*, pages 414–416. RESNA Press, 1997.
12. H.A. Yanco and J. Gips. Driver performance using single switch scanning with a powered wheelchair: robotic assisted control versus traditional control. In *Proceedings of the Rehabilitation Engineering Society of North America 1998 Annual Conference*. RESNA Press, 1998.
13. H.A. Yanco, A. Hazel, A. Peacock, S. Smith and H. Wintermute. Initial report on Wheelesley: a robotic wheelchair system. In *Proceedings of the IJCAI-95 Workshop on Developing AI Applications for the Disabled*, Montreal, Canada, August 1995.

Subject Index

Springer
and the
environment

At Springer we firmly believe that an international science publisher has a special obligation to the environment, and our corporate policies consistently reflect this conviction.

We also expect our business partners – paper mills, printers, packaging manufacturers, etc. – to commit themselves to using materials and production processes that do not harm the environment. The paper in this book is made from low- or no-chlorine pulp and is acid free, in conformance with international standards for paper permanency.

Springer

Lecture Notes in Artificial Intelligence (LNAI)

Lecture Notes in Computer Science